Biodiversity in Horticultural Crops

Biodiversity in Horticultural Crops

Brindra Chauhan

RANDOM

RANDOM PUBLICATIONS

NEW DELHI (INDIA)

Biodiversity in Horticultural Crops

ISBN 978-93-5111-771-1

Published in 2016 in India by

RANDOM PUBLICATIONS

4376-A/4B, Gali Murari Lal, Ansari Road
New Delhi-110 002
Phone : +9111-43580356, 011-23289044, 011-43142548
e-mail: sales@randompublications.com,
info@randompublications.com, randomexports@gmail.com

Reprinted 2025

Type Setting by : Friends Media, Delhi-110089
Printed at: Replika Press Pvt. Ltd.

Preface

Global Biodiversity is the variety of different types of life found on Earth and the variations within species. It is a measure of the variety of organisms present in different ecosystems. This can refer to genetic variation, ecosystem variation, or species variation within an area, biome, or planet. Terrestrial biodiversity tends to be highest near the equator, which seems to be the result of the warm climate and high primary productivity. Biodiversity is not distributed evenly on Earth. It is the richest in the tropics. Marine biodiversity tends to be highest along coasts in the Western Pacific, where sea surface temperature is highest and in the mid-latitudinal band in all oceans. There are latitudinal gradients in species diversity. Biodiversity generally tends to cluster in hotspots, and has been increasing through time but will be likely to slow in the future.

Horticulture is the branch of agriculture that deals with the art, science, technology, and business of vegetable garden plant growing. It includes the cultivation of medicinal plant, fruits, vegetables, nuts, seeds, herbs, sprouts, mushrooms, algae, flowers, seaweeds and non-food crops such as grass and ornamental trees and plants. It also includes plant conservation, landscape restoration, landscape and garden design, construction, and maintenance, and arboriculture. Inside agriculture, horticulture contrasts with extensive field farming as well as animal husbandry. Horticulturists apply their knowledge, skills, and technologies used to grow intensively produced plants for human food and non-food uses and for personal or social needs. Their work involves plant propagation and cultivation with the aim of improving plant growth, yields, quality, nutritional value, and resistance to insects, diseases, and environmental stresses. They work as gardeners, growers, therapists, designers, and technical advisors in the food and non-food sectors of horticulture.

– Author

Contents

1

Importance of Horticulture Sector

THE WORKFORCE CHALLENGE IN HORTICULTURE

Horticulture is the 3rd largest sector in agriculture behind meat and grains. Its gross value of production in 2008-2009 was $8.39 billion and the sector employs around 200,000 people. A document "Future Focus" (HAL 2008) produced for Horticulture Australia Limited in recent times showed that by 2020 current growth rate would deliver an extra $0.9 billion in profit. However there was potential for a further $2.45 billion in profit if the sector was prepared to keep up with and improve on the performance of its competitors. That same document commented that the sector had an inability to attract and retain human capability and skilled staff and strategies needed to be developed to increase productivity, attract skilled labour at all levels and develop/adopt labour saving devices. It further recorded that the sector needed to be attractive for young people to enter at all stages of the supply chain as providers of innovation services. The document was silent on how these outcomes might be achieved. Workforce is thus a critical issue for the horticultural sector and an analysis of the current workforce would provide a benchmark for future action.

THE HORTICULTURE WORKFORCE

For purpose of analysis the horticulture sector is classified into two subsectors, amenity and production with production further divided into fruit/ nuts and vegetables. The components. Amenity horticulture employs more than 80 per cent of the horticulture workforce (ie around 160,000) whereas the major export sector provides the rest. There are some gender inequities with the proportion of females being at best 30 per cent overall and significantly lower in trades and labourer categories in the amenity subsector.

Previous studies have considered education for the horticulture sector in Australia. These include, *inter alia*, McEvilly and Aldous (2010), Guisard and Kent (2009) and Rayner *et al.* (2009). This study builds on those papers. For the industry to move forward it is reasonable to expect that education and training would play a significant role. However on closer inspection the data suggest that horticulture in Australia is not well placed. The very low level of tertiary qualifications in the sector with only flower growers above 10 per cent and most categories below 5 per cent. There is a stronger performance in respect of vocational qualifications with amenity horticulture being significantly more qualified (45 per cent having VET qualifications) than other categories.

The question arises as to how these levels of education relate more widely. The Australian community has about 26 per cent of its constituency with a university degree whereas the agricultural sector has around 7.8 per cent and horticulture 4.6 per cent. In the US around 30 per cent of the community has a degree whilst 25 per cent of its agricultural sector also has degree level qualifications. These data suggest that the horticulture sector in Australia operates at an educational level below what might be considered adequate in modern Australia and well below at least one of its international competitors. That said, it is also clear that vocational education is valued and that the levels in the horticulture sector overall are comparable with levels in the community generally. However the numbers are distorted with higher levels in the amenity subsector and significantly lower levels in the production subsector where the prospects for increased productivity and markets are purported to be high.

An Educated and Trained Workforce for Horticulture

In order to address the perceived shortage of university trained people, it is important to understand the current graduate completion numbers and associated trends. Whilst it can be argued that the degree does not need to be in horticulture, there ought to be an expectation that the industry would want a large proportion of such graduates to have the relevant expertise. A perspective is provided by considering the trends over the last two decades or so. In the 1980s the agricultural college network provided horticulture qualifications at degree and associate degree levels.

Data show that at that time there were in excess of 150 graduates annually. With the Dawkins reforms in higher education in 1989 the colleges were gradually absorbed into the revamped university sector and programmes were rationalised. Courses disappeared where market forces did not support retention. Completions steadied to around 120 for the period up to 2005 but have been in steady decline since then. In 2010 there were fewer than 80 graduates in horticulture from Australian universities. It needs to be recognised that the vast majority of these graduates are amenity horticulturists and very few represent the production horticulture subsector and mainly in viticulture. As student intakes decline, universities respond by contraction since funding follows student load. Horticulture courses are no different and there has been significant loss of offerings across Australasia. In relatively recent times up to ten or so institutions offered a degree in horticulture in Australia and New Zealand. A survey in 2011 showed that only Charles Sturt University still offered a horticulture degree and the University of Melbourne offered an associate degree.

The University of Queensland, The University of Sydney and Massey University now offer a strand in another degree whilst The University of Western Sydney, the University of Adelaide, the University of Tasmania and Lincoln University offer one or more units of study. Ballarat University has a degree on its books but it is not offered. It is clear that access to tertiary horticultural education has been severely eroded as a result of lack of demand for such courses and there is no indication that the decline has stopped. Specialist viticulture programmes continue at Charles Sturt University, the University of Adelaide, Curtin University and Lincoln University but in most cases now as joint degrees with oenology. Offerings at La Trobe and the University of Western Australia have ceased. Viticulture is a component of the Wine Technology course at the University of Southern Queensland.

At the vocational level the number of completions were consistent for the period 2004-2007 being around 4500 per year nationally. Certificate 3 and Certificate 4 programmes were the most popular with low numbers at the higher level programmes. All states and territories provided courses with the majority in NSW followed by Queensland and Victoria.

Mathematical Connundrum

In order to maintain the education levels in the industry some benchmark calculations are warranted. The numbers of graduates required annually to maintain levels at known benchmarks. In the calculations it is estimated that the average career in horticulture is around 20 years. The numbers generated will increase with a shorter career span and decrease with a longer career span. In order to maintain the current level of 4.6 per cent university graduates in a workforce of 200,000 some 460 graduates per year are needed. If the horticulture industry had aspirations to match the levels in the agriculture sector or the general community then an annual supply of graduates needed would be 780 and 2,620 respectively without allowing for the graduate injection needed in the first place to be at the benchmark level. The annual supply is under 80 university graduates in horticulture and declining. This is under one fifth of that required to meet the current low levels of education. Clearly if the industry is to maintain its levels of university trained workforce, then graduates will have to come from other disciplines.

At the vocational (VET) level the situation is more positive. The horticultural industry has around 35 per cent of its workforce (ie about 70,000) with vocational qualifications. Assuming the 20 year career average, then a steady state operation would require annual additions of VET qualified people of around 3,500. That number equates to current rates of completions and is above the levels in the agricultural industry and in the community generally. It needs to be recognised that there is a significant proportion of VET training undertaken 'on the job' and so the calculations used simplify the situation somewhat but nevertheless provide a 'ballpark' estimate.

These calculations suggest that the horticultural industry faces a challenge at the managent level whereas training of the labour force seems well founded relative to other sectors of the economy.

The Need for Graduates

It is a reasonable question to ask why the industry needs more graduates. Doubters can point to the last century where the industry performed well and developed to current production levels. This was achieved without the emphasis on university qualifications but with reasonable labour force vocational training. Whilst that has been the case, its continuation unchanged ignores the realities of the new century. There will continue to be the need for strong vocational training but there is an ever increasing need for high level management and technical skills that ought not be ignored.

Modern horticultural operations have levels of complexity that did not exist even 20 years ago. The modern manager is faced with major compliance responsibilities such as occupational health and safety, pesticide management and environmental sustainability. Biosecurity is increasingly important. Business principles receive greater scrutiny with regular business activity statements. Marketing is increasingly the responsibility of the business, quality assurance is expected and accreditation will become more common. Climate change has to be managed and carbon emissions need to be minimised. The industry is becoming more 'high tech' with precision agriculture and computer technology applications increasingly available. If the industry is to move forward, compete with international colleagues and capitalise on opportunities then it will need to attract the best educated people available. Further, there will need to be a continued pipeline of 'experts' to supply inputs, including specialist advice, and manage supply chains to ensure the industry can deliver on the new paradigm.

Soderlund (2004) provided advice that the industry had to shift from "a market garden mentality and small business approach to a focus on supply chain management, technology (precision farming) and export competitiveness". He further advised that a significant limitation was accessing *suitably qualified professionals* in those disciplines. It appears that has been little movement in these directions in the intervening years. Continuing to ignore the realities will not serve the sector well.

Considering the Options

The most obvious option for remediation of professional horticulture is to increase the number of students studying horticulture at university. In reality this is problematical because unless there is an immediate and substantial increase in student numbers, horticulture as an option is likely to disappear from university offerings. The impact of this outcome is that the applied expertise will disappear as well as reduce R&D capability and postgraduate research training in applied areas.

The fallback position is to attract agricultural graduates to the horticulture sector. Such graduates have made strong contributions in the past. However the decline in availability of graduates in agriculture is similarly dramatic, falling from around 800 in the 1980s to under 300 in 2010. If graduates in agricultural economics, agribusiness, horticulture/viticulture and animal science are added then that graduate number increases to around 700. Data show that the job market for graduates has been consistently in excess of 4000 per year for the recent past and so there is an acute shortage of graduates across the industry.

The consequences are stark. There are now fewer campuses that offer agricultural or horticultural education. This is particularly so in regional Australia where the number has halved in the last 20 years. Student access is thus becoming more restricted and there is a reluctance to travel to distant and metropolitan campuses because of the cost and dislocation from community. The shortage of graduates causes a reduction in the availability of informed advice to producers and there has been a significant increase in the number of 'snake oil salesmen' and associated unproven products. There is likely to be less capacity for R&D especially in applied science.

HIGHER EXPORT EARNING POTENTIAL

Fruits and vegetables have been shown to earn 20-30 times more foreign exchange per unit area than cereals due to higher yields and higher price available in the international market. In spite of the large scale production, India's fresh produce has not made major dent in the international trade so far, except for spices, cashew nut and recently grapes. This is primarily due to the fact that horticultural crops were treated as one of the several means of land use of secondary importance, with food grain crops receiving prime attentions and consequently heavy investments.

PHYSIOLOGICAL AND BIOCHEMICAL ASPECTS

The quality of the harvested fruits and vegetables depend on the condition of growth as well as physiological and biochemical changes they undergo after harvest. Fruits and vegetable cells are still alive after harvest and continue their physiological activity.The post harvest quality and storage life of fruits appear to be controlled by the maturity. If the fruits are harvested at a proper stage of maturity the quality of the fruits is excellent. Poor quality and uneven ripening are due to early harvesting and late harvesting which results in extremely poor shelf life. Respiration plays a very significant role in the post harvest life of the fruits. In most of the fruits, the rate of respiration increases rapidly with ripening.

The sudden upsurge in respiration is called the 'climacteric rise', which is considered to be the turning point in the life of the fruit. After this the senescence and deterioration of the fruit begin. The fruits such as banana, papaya, mango, guava, jackfruit etc. belong to the category of climacteric fruits. While litchi, pineapple, grapes, pomegranate, lemon, orange, lime, etc. belong to the non-climacteric group. To extend the post harvest life of the fruits its respiration rate should be reduced as far as possible. Thus an understanding of the factors, which influence the rate of respiration, is indispensable to post harvest technologies for manipulating the storage behaviour of fruits.

Mechanical Injury

Owing to their tender texture and high moisture content, fresh fruits and vegetables are very susceptible to mechanical injury. Poor handling, unsuitable containers, improper packaging and transportation can easily cause bruising, cutting, breaking, impact wounding and other forms of injury.

Parasitic Diseases

High post-harvest losses are caused by the invasion of fungi, bacteria, insects and other organisms. Microorganisms attack fresh produce easily and spread quickly, because the produce does not have much of a natural defence mechanism and has plenty of nutrients and moisture to support microbial growth. Post harvest decay control is becoming a more difficult task, because the number of pesticides available is falling rapidly as consumer concern for food safety increases.

Sites of losses

Losses may occur anywhere from the point where the food has been harvested or gathered up to the point of consumption. For the sake of convenience the losses can be broken down into the following sub-headings:

- *Harvest:* The separation of the commodity from the plant that produced it. In the case of roots, tubers and bulbs the commodity is lifted out of the soil.
- Preparation The preliminary separation or extraction of the edible from the non-edible portion, *e.g.*, the peeling of fruits and vegetables.
- Preservation is the prevention of lose and spoilage of foods. For example, the sun-drying of fruit, the use of refrigeration and the use of funficides to inhibit mold growth in fruits.
- Processing is the conversion of edible food into another form more acceptable or more convenient to the consumer, for example, the manufacture of fruit juice and the canning of fruits and vegetables.

- Storage is the holding of foods until consumption. Most storage is common storage (ambient temperature) but there are extensive storage capacities that can hold food under refrigerated or controlled atmosphere conditions.

Transportation. All forms of transportation are used to convey foods from the point of production to the ultimate point of consumption..

To summarize, transport losses are due to the following:

- Unsuitable transport containers;
- Overloading of mixed fruits and vegetables (in some developing countries people and even animals ride on top of the load);
- Irresponsible driving;
- Lack of feeder roads loading to highways or collection centres;
- Rough roads;
- Heat accumulation or very poor ventilation within the transport vehicles;
- Virtual absence of refrigerated and insulated trucks;
- Delays in product procurement after harvesting or at collection centres.

In addition, where machine is used for harvesting losses may be fairly high because the agricultural machinery that is used to harvest the crops leaves some of the commodity in the field and mechanically damages some of it. Considerable quantities of foods a may be discarded at the point of harvest because they are of the wrong size, shape or colour.

These are planned losses. But in the case of Bihar and Uttar Pradesh are lower because most of the crop is hand picked. The amount of material rejected is also lower because the expectation of quality and uniformity is lower in the local market.

In developed countries losses are generally small during processing, storage and handling because of the efficiency of the equipment, good quality storage facilities, and close control of critical variables by a highly knowledgeable cadre of managers. In contrast, in India, losses in processing, storage and handling tend to be rather high because of poor facilities and frequently inadequate knowledge of methods to care for the food properly.

The deterioration of a product starts during the harvesting operation. The more carefully a product is handled, the slower the deterioration process during subsequent handling operations.

The deterioration of a product starts during the harvesting operation. The more carefully a product is handled, the slower the deterioration process during subsequent handling operations. However, the farmers may be unaware or indifferent to the condition of the product after harvest, and harvesting procedures may thus be rather careless. The only constraint is to avoid external injury. Until farmers are convinced that careful handling will increase profits, it will be difficult to persuade them otherwise. Grading is a thing that most farmers are loath to do until they are convinced it will bring them added revenues or increases the acceptance of their products. A farmer may separate different varieties if they are distinguishable, but consumers of both states usually are more price conscious than quality conscious. That is not to say that they are unmindful of the quality, but if good quality products are too highly priced, they are often willing to settle for a poorer quality product.

Post harvest Technologies

Precooling Good temperature management is the most effective way to reduce post-harvest losses and preserve the quality of fruits and vegetables. Products harvested from hot fields often carry field heat and have high rates of respiration. Rapid removal of field heat by precooling is so effective in quality preservation that this procedure is widely used for highly perishable fruits and vegetables. Currently used precooling methods include room cooling, forced-air cooling, water cooling, vacuum cooling and package icing

Room cooling is a relatively simple method, which needs only a refrigerated room with adequate cooling capacity. The produce is packed in containers, which are loosely stacked in the cooling room, leaving enough space between containers for each one to be exposed to circulating cold air. The rate of cooling is rather slow compared to other methods of cooling, because the heat inside each container needs to be transferred to the surface of the container by means of conduction before being carried away by the refrigerated air. It may take hours or even days to cool a product, depending on what kind of product it is, the size and nature of the container, and the temperature and velocity of the circulating air.

Forced-air cooling is a more rapid way of using air to cool produce. Cold air is forced to flow through the inside of each container, so that it carries away heat directly from the surface of the produce rather than from the surface of the container. The airflow is produced by creating a pressure difference between the two perforated sides of each container. The containers are stacked inside a covered tunnel with an exhaust fan at one end. Highly perishable and high-value products such as grapes, strawberries and raspberries may be cooled in less than an hour using this method.

Hydro cooling is a rapid and less expensive method. Produce is exposed to cold water by means of showering or dipping. The required cooling time is often a matter of minutes. However, not all kinds of products tolerate hydro cooling. Hydro cooled products inevitably have a wet surface, which may encourage decay in some kinds of produce.

Vacuum cooling is the most efficient way to cool leafy vegetables, particularly headed ones such as head lettuce, cabbage and Chinese cabbage. The produce is placed inside a vacuum tube in which air pressure is reduced.

When the pressure is lowered to 4.6 mm Hg, water "boils" off at o 0°C from all over the leaf surface. The boiling effect draws heat for vaporization, and hence cools the produce. The cooling time is usually in the order of 20 to 30 minutes. Unfortunately, the equipment needed for vacuum cooling is very expensive, and may not be a good choice for small-scale farming systems.

Ice bank cooler is a new development in refrigeration with positive ventilation. In this system ice cool air is passed through the boxes containing horticultural produce. This facilitates quicker cooling and large amount of heat is removed in a relatively shorter period. The store maintains a temperature of 0.5 - 0.8oC and relative humidity of 98 per cent.

Package icing or top icing is the simplest way of cooling. Adding crushed ice, flake-ice or slurry of ice in containers can cool the produce. However, this method is not suitable for produce, which is very sensitive to ice-cold temperatures. Cooling by ice also inevitably wets both the produce and container, and generates water, which needs to be drained.

Sanitation

Sanitation is of great concern to produce handlers, not only to protect produce against post harvest diseases, but also to protect consumers from food borne illnesses.

E.coli 157:H7, *Salmonella, Chryptosporidium, Hepatitis and Cyclospera* are among the disease causing organisms that have been transferred via fresh fruits and vegetables. Use of a disinfectant in wash water can help to prevent both post harvest diseases and food borne illnesses.

Chlorine in the form of a sodium hypochlorite solution or as a dry powdered calcium hypochlorite can be used in hydro-cooling or wash water as a disinfectant. For the majority of vegetables, chlorine in wash water should be maintained in the range of 75-150 pm (parts per million). The antimicrobial form, hypochlorous acid, is mostly available in water with a neutral pH (6.5 to 7.5). Organic growers must use chlorine with caution, as it is classified as a restricted material.

Ozonation is another technology that can be used to sanitize produce. A naturally occurring molecule, ozone is a powerful disinfectant. Fruit and vegetable growers have begun using it in dump tanks as well, where it can be thousands of times more effective than chlorine. Ozone not only kills whatever food borne pathogens might be present, it also destroys microbes responsible for spoilage. A basic system consists of an ozone generator, a monitor to gauge and adjust the levels of ozone being produced and a device to dissolve the ozone gas into the water.

Hydrogen peroxide can also be used as a disinfectant. Concentrations of 0.5 per cent or less are effective for inhibiting development of post harvest decay caused by a number of fungi. Hydrogen peroxide has a low toxicity rating and is generally recognized as having little potential for environmental damage.

Presizing and Storage

For many commodities fruits below a certain size are eliminated manually or mechanically by presizing belt. Undersized fruits are diverted for processing. The sorting process eliminates cull, overripe, misshapen and otherwise defective fruit and separates produce by colour, maturity and ripeness classes.

Grading

Essentially all fruits and vegetables sold in modern markets are graded and sized into two or more grades according to trade standards. Sophisticated marketing systems require precise grading standards for each kind of product. More primitive markets may not use written grade standards, but the products are sorted and sized to some extent.

Typical grading facilities in large packinghouses include dumpers and conveyors. Produce is graded by human eyes and hands while moving along conveyor belts or rollers. "Electric eyes" are sometimes used to sort produce by colour. In small scale packing operations, one or a few grading tables may be enough. Dumping, conveying and grasping can cause mechanical injury to some products. Equipment should have a smooth, soft surface and dumping and grading operations should be gentle to minimize injuries.

Many products are sized according to their weight. Automated weight sizers of various capacities are used in packinghouses. Round or nearly round fruits are often sized according to their diameter, using automated chain or roller sizers or hand carried ring sizers. An inefficient sizing operation can also cause significant injuries.

Waxing

Food grade waxes are commonly applied to replace some of the natural waxes removed in the washing and cleaning operations to reduce water loss and to improve appearance. It also provides protection against decay organisms. Waxing may be done after grading and fungicides may be added to the wax. Application of wax and post harvest fungicides must be indicated on each container where the refrigerated storage facilities are not available Protective skin coating with wax is one of the methods for increasing the storage life of fresh fruits.

Packaging

Packaging of fresh fruits and vegetables has a great significance in reducing the wastage. Packaging provides protection from physical damage during storage, transportation and marketing. There are variety of packages, packaging materials and inserts available. There are two types of packaging. The first is when produce is packed in containers for transportation and wholesale.

The second is when produce is packed into small retail units. Ideal containers for packing fruits and vegetables should have the following attributes. They are easy to handle, they provide good protection from mechanical damage, they have adequate ventilation and they are convenient for merchandising. They should also be inexpensive and easily degradable or recyclable. Many kinds of containers have been used but the "ideal" is yet to be found. Users often put economic considerations first is selecting containers. Fancy containers such as fibreboard boxes or wooden or plastic crates, are often used for high-value products. Inexpensive containers such as bamboo baskets or nylon net sacs are used for low-priced produce. Methods of packaging can affect the stability of products in the container during shipping, and influence how much the container protects their quality. In fibreboard boxes, for example, delicate and high-priced products are often packed in trays, while other products are simply put in the box in groups.

Prepackaging or consumer packaging generally provides additional protection for the products. It is also convenient for retailers as well as customers, and therefore adds value to produce. However, over-use of non-biodegradable plastic trays and wrapping materials, as often seen in modern supermarkets, which creates an extra burden of waste disposal and damages the environment.

HORTICULTURE SCORES IN EMPLOYMENT GENERATION

The average labour requirement for fruit production is 860 man days per hectare per annum as against 143 man days for cereal crops. Crops like grapes, bananas and pineapple generate much larger employment roughly from 1000-2500 man days per hectare per annum.

The vast opportunities for investment in fruits and vegetable processing units exist. It is established that fresh agri-produce loss reduction is cheaper than equivalent increase in production so far as economy, energy and impact on environment is concerned. Theoretically, one per cent post-harvest loss reduction of horticulture produce is expected to save Rs 230 crores annually.

If the consumption level shoots up from the current 100 gm of fruit and 200 gm of vegetables per capita per day to at least the recommended dietary level of 140 gm and 270 gm respectively by 2010, the domestic market for fresh fruits and vegetables could be as large as Rs 50,000 crores at today's price structure.

HARVEST HANDLING OF HORTICULTURE PRODUCE

As far as state of the horticulture sector of Uttar Pradesh is concerned, it has many advantages over Bihar in term of infrastructure and access to the big market for its horticulture produce. As far as power condition is concerned, situation is not much different from Bihar. Uttar Pradesh is also facing acute power shortage. This is certainly adversely affecting cold storage like other industrial units. Though, recently some steps have been taken to ensure regular power supply to cold storages, nonetheless,losses of a portion of total horticulture produce can not be ruled out. According to our estimate, it may vary between 5 to 35 per cent of the total produce. In case of Uttar Pradesh, in addition to the existence of a big market in UP itself, its closeness to Delhi, has given a big boost to its horticulture produce. As a result export of fruits and other horticulture produce have also become feasible.

In spite of so many positive factors, horticulture sector of Uttar Pradesh is still far from realisation of its actual potential. Many incentives are required on behalf of government to induce farmers to think to switch over to horticulture sector on a large scale.

First and foremost impediment is the market. Regional and local markets are small in size and they do not have the capacity to absorb any horticulture produce in large quantity. Gone the days when Indian agriculture sector was facing the problem of undercapacity or under production. Now, Indian agriculture sector is faring well but result is more disastrous than before. It is driving farmers to suicide. Indian horticulture sector is also not an exception of this trend.

Almost same thing is happening also in this sector. For instance, take example of bumper production of potato in Uttar Pradesh. This year's estimated 4 – 4.5 million excess production of potato is not contributing to the betterment of farmers, it actually making them economically more disadvantaged. Therefore, without a market, it is naïve to suggest farmers to resort to horticulture produces.

Recently, some training programmes for pre and post harvest management for horticulture produce are being undertaken with the assistance from APEDA, National Horticulture Board and from National Institute of Agricultural Marketing (NIAM). But this has to be taken up at a large scale, only then situation can improve substantially.

Road conditions and communication facilities need a lot of improvement. Common farmer does not know how to reach Paris or Dubai markets? So, market intelligence is a major bottleneck. They can produce good quality fruits and vegetables but where is their market? So, arrangement should be worked out so that market intelligence can be provided to the farmers at their doorsteps.

Everyone is harping on diversification of agriculture but one would like to know that what kinds of incentive are given by the state government or any agency for rapid diversification of agriculture – nothing. Therefore, some frsh incentives are required.

In eastern Uttar Pradesh, sugar cane used to be a main cash crop for the farmer. Since sugar industry is in a very bad shape, most of the farmers want to switch over to another cash crops, possibly in horticulture. But one prime concern of a farmer is the market. Where is the market for horticulture produce? Bumper production of potato is actually playing havoc with the farmers livelihood. Many cases have been noticed when farmers keep their potato in cold storage but they do not come to take it back. Reason is prevailing low price of potato in the market. That price actually does not cover the cost of transportation and cold storage charges. By June 30 per cent of Potato, used to be taken out of the cold storage but this year by last week of June even 5 per cent of Potato has not been taken out of cold storages. Reason is, the price farmers have to pay for per quintal Potato to keep their produce in cold storage is almost equal to its prevailing price in the whole sale market of Potato. As a result, cold storage owners fear for the worst. They are afraid that the farmers will not come back to take back their produce and ultimately they will have to arrange buyers for it.

As far as state of the other horticulture produce is concerned, situation is even worse than this. Cold storage has no meaning for horticulture produce other than Potato because the temperature tthat require for other produce is not available in these cold storages. Even they don't have different kind of chambers for different produce. Therefore, all kinds of produce are kept in the same chambers, that also create problems because various horticulture produce required different temperature but they are given the same temperature. Therefore, neither it is advisable nor desirable for the farmers- given the present state of infrastructure and market conditions-to switch over from other cash crops to the horticulture.

Implementation of proper post-harvest management methods in the horticulture industry is expected to limit losses to less than 5 per cent in these districts. Because, in the case of other horticulturally developed countries like China, Korea and Japan, the level of post-harvest losses is below 5 per cent.

Lack of advanced infrastructure facilities such as sorting and grading of products, transportation and marketing system, cold storage facilities are yielding very low returns as a result farmers are reluctant to make further investment in the industry. And this phenomena is not limited to a particular district or state, it is a national phenomena.

From national viewpoint, even though India enjoys an enviable position in the production of horticultural crops, the productivity level is much lower than other developed countries owing to non-availability of quality plant and seed material. Not only this, there is also little chance of getting technical assistance in the field of fruits and vegetable cultivation in India as there is no strong network between the horticulturally developed countries and India. Presently, low cost processing technologies are not available to Indian farmers, as a result only less than 0.1 per cent of the entire production is being processed. Whereas, the potential demand for processed fruits and vegetables is at 0.5 per cent, with 55 per cent in the export and 45 per cent in the domestic market.

To promote the horticulture industry in the country especially through proper post-harvest strategies in areas such as sorting and grading, packing, transportation, storage, infrastructure development, several initiatives are needed to be taken, and some steps have already been taken. Interest-free loans are being offered to farmers for setting up grading and packing centres, retail outlets, pre-cooling units, purchase of plastic crates and also transport vehicles. The assistance is being given in the form of soft loan with 4 per cent service charges and better repayment schedule at competitive rate of interest.

HARVEST HANDLING OF HORTICULTURE PRODUCE

Insufficient allocation of money in proportion to its contribution to GDP and generation of employment opportunities, is reflected in inadequate infrastructure for horticulture produce. Initial indication of this survey points towards complete negligence of horticultural infrastructure facilities in some districts of Bihar, while in other parts, situation is comparatively better. It was found that, while there is only one cold storage in Saharsa, that cater the necessity of all neighbouring districts' horticulture producers. Though it caters the necessity of local horticulture producers to some extent, it can not fulfil the necessity of far - flung rural areas of even saharsa district, forget necessity of other adjoining districts.

Another important point regarding horticulture sector of this region has been noticed that main purpose behind production is not a commercial one but self-consumption. Main reason behind this mindset is their traditional bias towards other crops. Generally, a particular community called 'Kujra' only produce vegetables for commercial purpose. Level of low commercial awareness among farmers about the importance of horticulture produce is also one of the reason.

When surveyors went to the villages and asked about cold storage facility, some of them were not even aware of that existing facility. But some of the farmers who were aware of it, but were not willing to use this facility because only cold storage of that area was situated in the city centre and distance between the city centre and the village (production centre of the horticulture produce), given the bad transportation facilities, actually made it unattractive for the horticulture producers to keep their produce in cold storage. Even farmers of saharsa district, on average will have to carry their produce from 10 to 50 kilometers to keep their horticulture produce in that cold storage. If farmers of the neighbourhood districts decide to use the same facilities, on average they will have to carry it from 30 to 125 kilometers. And all this will only add to the cost of their produce and that will ultimately result in either low profit or no profit at all.

Lack of capital also force farmers to ignore the use of cold storage, even when available and effectively managed. Many growers depend on almost daily sales for their income and hence may be forced to accept a lower price immediately, rather than to store their produce in the anticipation of a higher price. There is also the storage rental price which the farmer may not be willing to pay unless he is thoroughly convinced that he will not only recover his investment, but will also profit.

Another problem is the lack of access to the big market. Since most of the horticulture produce of these areas, do not have direct access to the big cities, as a result either they produce for self-consumption or for local market, size of which market is very small. As a result, only a fraction of farmers are involved in commercial production of fruits and vegetables.

Main horticulture produce of these districts are Mango, Litchi, Banana, Guava, Brinjal, Potato, Cabbage, Tomato, Lady finger, Cauliflower, Parval etc.

Our researchers found that sometimes some horticulture produce, because of its perishable nature and lack of storage and processing facilities, are being sold below cost of its production. According to our estimate, total loss of horticulture produce in these districts may vary between 25 –33 per cent of total horticulture produce, due to improper handling of harvest and post-harvest management techniques resulting in a loss of thousands of crore every year. Though, post – harvest handling of horticulture produce is better in Muzaffarpur, Begusarai and Hajipur, still a lot of scope for value addition and processing exist. According to our casual estimate for these districts, it may vary between 25 – 30 per cent of total horticulture produce.

As far as state of the horticulture produce in other and comparatively developed districts of Bihar is concerned, over the years losses have been reduced sharply because of phenomenal increase in total number of cold storage and processing units as well. Though, barring few, most of the processing units are of small size. They have advantage of a big nearby market of Patna and of comparatively better transportation facilities as well. But frequent power cuts and loadshedding is a cause of concern for them.

CONCLUSION

1. Horticultural crops not only provide us with nutritional and healthy foods, but also generate a cash income to growers. It has full potential to become an alternate cash crop for sugarcane grower farmers of Uttar Pradesh and Bihar.

On the other hand, all countries are veering towards self-reliance in foodgrains. This trend may well augur the gradual diminishing of foodgrain exports. On the other hand, the demand for fruits and vegetables is steadily increasing. And India has the unique distinction of being able to grow almost all fruits and vegetables. However, export-oriented production is still at the planning stage: most Indian producers have yet to adopt world standards of quality. So far as its contribution to Indian economy is concerned, fruits and vegetables are grown only on 7 – 8 per cent of gross cropped area but contribute more than 18.8 per cent of the gross value of agricultural output and 52 per cent export earnings from total agricultural produce. During the last few years, considerable emphasis has been given to this sector: Accordingly areas under fruit production increased by 172 per cent from 1961 - 1993, productivity per hectare was nearly doubled, leading to an increase in production to the tune of 320 per cent.

Though, huge employment and export potential of horticulture has been fully recognised by the successive governments, still it is not getting the kind of priority it actually deserves. As the Tenth Plan draft acknowledges that the ' horticulture sector contributes about 24.5 per cent towards agriculture GDP from only about 8 per cent of the cultivated area'. It further says, ' besides, providing nutritional and livelihood security and helping poverty alleviation and employment generation, this sub-sector sustains a large number of agro – industries, which generate huge additional non –farming employment opportunities'.

Despite all these admissions, in practice while ₹ 22467 crore was allocated to agricultural sector in Eighth Plan, only ₹ 1000 crore was earmarked, out of that for horticulture. In Ninth Plan also situation could change only marginally. While total allocation to agriculture sector was almost doubled from Rs.22467 to ₹ 42462 crore, out of which amount allocated to horticulture sector was raised from ₹ 1000 crore to meagre ₹ 1400 crore

2

Organic Horticulture

Organic horticulture is the science and art of growing fruits, vegetables, flowers, or ornamental plants by following the essential principles of organic agriculture in soil building and conservation, pest management, and heirloom variety preservation.

The Latin words hortus and cultura together form horticulture, classically defined as the culture or growing of garden plants. Horticulture is also sometimes defined simply as "agriculture minus the plough." Instead of the plough, horticulture makes use of human labour and gardener's hand tools, although some small machine tools like rotary tillers are commonly employed now.

GENERAL

Double Digging, Vermicompost, Mulches, cover crops, compost, manures, and mineral supplements are soil-building mainstays that distinguish this type of farming from its commercial counterpart. Through attention to good healthy soil condition, it is expected that insect, fungal, or other problems that sometimes plague plants can be minimized. However, pheromone traps, insecticidal soap sprays, and other pest-control methods available to organic farmers are also sometimes utilized by organic horticulturists.

Horticulture involves five areas of study. These areas are floriculture, landscape horticulture, olericulture and postharvest physiology. All of these can be, and sometimes are, pursued according to the principles of organic cultivation. Organic horticulture is based on knowledge and techniques gathered over thousands of years. In general terms, organic horticulture involves natural processes, often taking place over extended periods of time, and a sustainable, holistic approach - while chemical-based horticulture focuses on immediate, isolated effects and reductionist strategies.

PEST CONTROL APPROACHES

Differing approaches to pest control are equally notable. In chemical horticulture, a specific insecticide may be applied to quickly kill off a particular insect pest. Chemical controls can dramatically reduce pest populations in the short term, yet by unavoidably killing natural control insects and animals, cause an increase in the pest population in the long term, thereby creating an ever increasing problem. Repeated use of insecticides and herbicides also encourages rapid natural selection of resistant insects, plants and other organisms, necessitating increased use, or requiring new, more powerful controls.In contrast, organic horticulture tends to tolerate some pest populations while taking the long view.

Organic pest control requires a thorough understanding of pest life cycles and interactions, and involves the cumulative effect of many techniques, including:

- Allowing for an acceptable level of pest damage
- Encouraging predatory beneficial insects to flourish and eat pests
- Encouraging beneficial microorganisms
- Careful plant selection, choosing disease-resistant varieties
- Planting companion crops that discourage or divert pests
- Using row covers to protect crop plants during pest migration periods
- Rotating crops to different locations from year to year to interrupt pest reproduction cycles
- Using insect traps to monitor and control insect populations

Each of these techniques also provides other benefits, such as soil protection and improvement, fertilization, pollination, water conservation and season extension. These benefits are both complementary and cumulative in overall effect on site health. Organic pest control and biological pest control can be used as part of integrated pest management. However, IPM can include the use of chemical pesticides that are not part of organic or biological techniques.

FOREST GARDENING

A forest garden is a designed agronomic system based on trees, shrubs and perennial plants. These are mixed in such a way as to mimic the structure of a natural forest - the most stable and sustainable type of ecosystem in this climate.

The primary aims for the system are:

- To be biologically sustainable, able to cope with disturbances such as climate change
- It should be productive, yielding a number of different products
- It should require low maintenance.

The crops which are produced will often include fruits, nuts, edible leaves, spices, medicinal plant products, poles, fibres for tying, basketry materials, honey, fuelwood, fodder, mulches, game, sap products.

Forest gardens have been used for millennia in tropical regions, where they still often form a major part of the food producing systems which people rely on, even if they work elsewhere for much of the time. They may also provide useful sources of extra income. Their use is intimately linked with prevailing socio-economic conditions. They are usually small in area, often 0.1-1 hectares. In temperate regions, forest gardens are a more recent innovation, many inspired by Robert Hart's efforts in Shropshire over the last 30 years. A major limiting factor for temperate forest gardens in the amount of sunlight available to the lower layers of the garden: in tropical regions, the strong light conditions allow even understorey layers to receive substantial light, whereas in temperate regions this is not usually the case. To compensate for this, understorey layers in temperate forest gardens must be chosen very carefully - there are plenty of plant crops which tolerate shady conditions, but many are not well known. Many of the more common shrub or perennial crops need bright conditions, and it may be necessary to design in more open clearings or glades for such species. Temperate forest gardens are also usually small in area, from tiny back garden areas up to a hectare in size. While food production and land use remain the concern of a minority of landowners and businesses, their use is likely to be limited to 'alternative' and organic gardeners and land users.

The key features which contribute to the stability and self-sustaining nature of this system are:

- The large number of species used, giving great diversity
- The careful inclusion of plants which increase fertility, such as nitrogen fixers.
- The use of dynamic accumulators - deep rooting plants which can tap mineral sources deep in the subsoil and raise them into the topsoil layer where they become available to other plants, eg. Coltsfoot, Comfreys, Liquorice, Sorrel.
- The use of plants specially chosen for their ability to attract predators of common pests, eg umbellifers like tansy.
- The use, where possible, of pest and disease resistant varieties, eg. apples.
- The increasing role of tree cover and leaf litter which improve nutrient cycling and drought resistance.

A forest garden is organised in up to seven 'layers'. Within these, the positioning of species depends on many variables, including their requirements for shelter, light, moisture, good/bad companions, mineral requirements, pollination, pest-protection, etc.

The layers consist of:

- *Canopy trees*: The highest layer of trees. May include species such as Chestnuts, Persimmons, honey locusts, Strawberry trees, Siberian pea trees Cornelian cherries, Azeroles and other hawthorn family fruits, Quinces, Apples, Medlars, Mulberries, Plums, Pears, highbush cranberries.

- Small trees and large shrubs, mostly planted between and below the canopy trees. May includes some of the canopy species on dwarfing rootstocks, and others such as various bamboos, Serviceberries, Plum yews, Chinkapins, Elaeagnus spp, and Japanese peppers. Others may be trees which will be coppiced to keep them shrubby, like medicinal Eucalyptus spp, and beech and limes with edible leaves.
- Shrubs, mostly quite shade tolerant. May include common species like currants and berries, plus others like chokeberries, barberries, Chinese dogwood, Oregon grapes, New Zealand flax and Japanese bitter oranges.
- Herbaceous perennials, several of which are herbs and will also contribute to the ground cover layer by self-seeding or spreading. These may include Bellflowers with edible leaves, Comfreys, Balm, Mints, Sage, and Tansy.
- Ground covers, mostly creeping carpeting plants which will form a living mulch for the 'forest floor'. Some may be herbaceous perennials, others include wild gingers, cornels, Gaultheria spp, and carpeting brambles.
- *Climbers and vines*: These are generally late additions to the garden, since they obviously need sturdy trees to climb up. They may include hardy kiwis and grapes.
- The final 'layer' is the root zone or rhizosphere. Any design should take account of different rooting habits and requirements of different species, even if root crops are not grown much. Some perennials with useful roots include liquorice and the barberries whose roots furnish a good dye and medicinal products. Various beneficial fungi can also be introduced into this layer.

VEGAN ORGANIC GARDENING

Vegan organic gardening and farming is the organic cultivation and production of food crops and other crops with a minimal amount of exploitation or harm to any animal. Vegan and vegan-organic farmers use no animal products or by-products, such as bloodmeal, fish products, bone meal, feces, or other animal-origin matter, because they view the production of these materials as either harming animals directly, or as being associated with the exploitation and consequent suffering of animals.

Some of these materials are by-products of animal husbandry, created during the process of cultivating animals for the production of meat, milk, skins, furs, entertainment, labour, or companionship; the sale of by-products decreases expenses and increases profit for those engaged in animal husbandry, and therefore helps support the animal husbandry industry, an outcome most vegans find unacceptable.

Soil fertility is maintained by the use of green manures and composted vegetable matter and minerals. Some vegan gardeners supplement this with human urine and 'humanure' produced from compost toilets; others avoid the potential health risks of using human waste. Such wastes may technically be considered 'animal products', however many vegan organic growers do not consider their usage unacceptable as there is unlikely to have been exploitation associated with their production.

VEGAN ORGANIC GARDENING

Many vegan organic gardeners prepare soil for cultivation using the same time-honoured method used by conventional and organic gardeners of breaking up the soil with hand tools and power tools and allowing the weeds to decompose. Shallow cultivation is also becoming popular among such gardeners: shallowly turning the soil's surface using one of numerous surface cultivating tools that are now available in the market place, including the popular Coleman surface-hoes developed by Eliot Coleman. Shallow tilling disturbs the soil less than deep turning and, combined with avoidance of soil compaction as described by O'Brien helps the soil maintain ecological balance. This minimizes the opportunities for soil diseases, plant diseases, and insect pests to become abundant.

O'BRIEN VEGANIC GARDENING METHOD

The Kenneth Dalziel O'Brien Veganic Gardening Method is a distinct system that was developed by Rosa Dalziell O'Brien, Kenneth Dalziel O'Brien, and May E Bruce, although the term was originally coined by Geoffrey Rudd as a contraction of 'vegetable organic' in order to "denote a clear distinction between conventional chemical based systems and organic ones based on animal manures". The O'Brien system's principal argument is that animal manures are harmful to soil health rather than that their use involves exploitation of and cruelty to animals.

The Kenneth Dalziel O'Brien system employs very specific techniques including the addition of straw and other vegetable wastes to the soil in order to maintain soil fertility. Gardeners following the Dalziel O'Brien system use soil-covering mulches, and employ non-compacting surface cultivation techniques using any short-handled, wide-bladed, hand hoe.

They kneel when surface cultivating, placing a board under their knees to spread out the pressure, and prevent soil compaction. Kenneth Dalziel O'Brien published a description of his system in Veganic Gardening, the Alternative System for Healthier Crops.

From Veganic Gardening by Kenneth Dalziel O'Brien: "the veganic method of clearing heavily infested land is to take advantage of a plant's tendencies to move its roots nearer to the soil's surface when it is deprived of light. To make use of this principle, aided by a decaying process of the top growth of weeds, etc., it is necessary to subject such growth to heat and moisture in order to speed up the decay, and this is done by applying lime, then a heavy straw cover, and then the herbal compost activator.... The following are required: Sufficient new straw to cover an area to be leared to a depth of 3 to 4 inches..."

Also part of the O'Brien method is minimal disturbance of the soil by tilling, use of compost, mulch, cover crops, and green manures, use of permanent raised beds and permanent hard-packed paths between them, laying out beds from north to south, putting plants in double rows or more so that not every row has a path on both sides.

ORGANIC AGRICULTURE AND YIELD

Yields relative to comparable conventional systems are directly related to the intensity of farming of the prevailing conventional systems. This is not only the case for comparison between regions, but also between crops within a region, and for individual crops over time8.

An over-simplification of the impact of conversion to organic agriculture on yield indicates that:

- In intensive farming systems, organic agriculture decreases yield; the range depends on the intensity of external input use before conversion9.
- In the so-called green revolution areas, conversion to organic agriculture usually leads to almost identical yields.
- In traditional rain-fed agriculture, organic agriculture has shown the potential to increase yields.

A number of studies have shown that under drought conditions, crops in organic agriculture systems produce significantly higher yields than comparable conventional agricultural crops, often out-yielding conventional crops10 by 7–90 per cent. Others have shown that organic systems have less long-term yield variability.

A survey of 208 projects in developing tropical countries in which contemporary organic practices were introduced, showed average yield increases of 5–10 per cent in irrigated crops and 50–100 per cent in rainfed crops. The so-called organic transition effect, in which a yield decline in the first 1–4 years of transition to organic agriculture, followed by a yield increase when soils have developed adequate biological activity has not been borne out in some reviews of yield comparison studies.

Trials conducted on organic cotton at Nagpur indicated that after the third year, the organic plot, which did not receive fertilizers and insecticides, produced as much cotton as that cultivated with them. Similarly, studies conducted in Punjab clearly indicated that organic farming gave higher or equal yields of different cropping systems compared to chemical farming after an initial period of three years.

PROGRESS OF ORGANIC FARMING

The most important constraint felt in the progress of organic farming is the inability of the government policy making level to take a firm decision to promote organic agriculture. Unless such a clear and unambiguous direction is available in terms of both financial and technical supports, from the Centre to the Panchayath levels, mere regulation making will amount to nothing. The following are found to be the major problem areas for the growth of organic farming in the country:

It is a fact that many farmers in the country have only vague ideas about organic farming and its advantages as against the conventional farming methods. Use of bio-fertilizers and bio pesticides requires awareness and willingness on the part of the farming community. Knowledge about the availability and usefulness of supplementary nutrients to enrich the soil is also vital to increase productivity.

Farmers lack knowledge of compost making using the modern techniques and also its application. The maximum they do is making a pit and fill it with small quantities of wastes. Often the pit is flooded with rainwater and result is the top of the compost remains under composted the bottom becomes like a hard cake. Proper training to the farmers will be necessary to make vermi-compost on the modern lines.

Attention on the application of composts/organic manure is also lacking. The organic matter is spread during the months when the right moisture level is absent on the soil. The whole manure turns into wastes in the process. The required operation is of course labour intensive and costly, but it is necessary to obtain the desired results.

It is found that before the beginning of the cultivation of organic crops, their marketability and that too at a premium over the conventional produce has to be assured. Inability to obtain a premium price, at least during the period required to achieve the productivity levels of the conventional crop will be a setback. It was found that the farmers of organic wheat in Rajasthan got lower prices than those of the conventional wheat. The cost of marketing of both types of products was also same and the buyers of wheat were not prepared to pay higher prices to the organic variety. Many experts and well informed farmers are not sure whether all the nutrients with the required quantities can be made available by the organic materials. Even if this problem can be surmounted, they are of the view that the available organic matter is not simply enough to meet the requirements.

The crop residues useful to prepare vermi-compost are removed after harvest from the farms.and they are used as fodder and fuel. Even if some are left out on the farms termites, etc destroy them. Experiments have shown that the crop residues ploughed back into soil will increase productivity and a better alternative is conversion into compost. The small and marginal cultivators have difficulties in getting the organic manures compared to the chemical fertilizers, which can be bought easily, of course if they have the financial ability. But they have to either produce the organic manures by utilizing the bio-mass they have or they have to be collected from the locality with a minimum effort and cost. Increasing pressure of population and the disappearance of the common lands including the wastes and government lands make the task difficult.

In spite of the adoption of the NPOP during 2000, the state governments are yet to formulate policies and a credible mechanism to implement them. There are only four agencies for accreditation and their expertise is limited to fruits and vegetables, tea, coffee and spices. The certiiying agencies are inadequate, the recognized green markets are non-existent, the trade channels are yet to be formed and the infrastructure facilities for verification leading to certification of the farms are inadequate. The small and marginal farmers in India have been practicing a sort of organic farming in the form of the traditional farming system. They use local or own farm renewable resources and carry on the agricultural practices in an ecologically friendly environment. However, now the costs of the organic inputs are higher than those of industrially produced chemical fertilizers and pesticides including other inputs used in the conventional farming system.

The groundnut cake, neem seed and cake, vermi-compost, silt, cow dung, other manures, etc. applied as organic manure are increasingly becoming costly making them unaffordable to the small cultivators.

Marketing Problems of Organic Inputs

Bio-fertilizers and bio-pesticides are yet to become popular in the country. There is a lack of marketing and distribution network for them because the retailers are not interested to deal in these products, as the demand is low. The erratic supplies and the low level of awareness of the cultivators also add to the problem. Higher margins of profit for chemical fertilizers and pesticides for retailing, heavy advertisement campaigns by the manufacturers and dealers are other major problems affecting the markets for organic inputs in India. Promotion of organic agriculture both for export and domestic consumption, the requirements of food security for millions of the poor, national self-sufficiency in food production, product and input supplies, etc. are vital issues which will have to be dealt with in an appropriate agriculture policy of India. These are serious issues the solution for which hard and consistent efforts along with a national consensus will be essential to go forward. Formulation of an appropriate agriculture policy taking care of these complexities is essential to promote organic agriculture in a big way.

The developing countries like India have to design a plethora of national and regional standards in attune with those of the developed countries. The adoption and maintenance of such a regulatory framework and its implementation will be costly. The cost of certification, a major component of which is the periodical inspections carried out by the certifying agencies, which have freedom to fix the timings, type and number of such inspections appears to be burdensome for the small and marginal farmers. Of course, the fees charged by the international agencies working in India before the NPOP were prohibitive and that was a reason for the weak response to organic agriculture even among the large farms in the country. No financial support as being provided in advanced countries like Germany is available in India. Supports for the marketing of the organic products are also not forthcoming neither from the State nor from the Union governments. Even the financial assistance extended to the conventional farming methods are absent for the promotion of organic farming.

In many cases the farmers experience some loss in yields on discarding synthetic inputs on conversion of their farming method from conventional to organic. Restoration of full biological activity in terms of growth of beneficial insect populations, nitrogen fixation from legumes, pest suppression and fertility problems will take some time and the reduction in the yield rates is the result in the interregnum. It may also be possible that it will take years to make organic production possible on the farm. Small and marginal farmers cannot take the risk of low jaelds for the initial 2-3 years on the conversion to organic farming. There are no plans to compensate them during the gestation period. The price premiums on the organic products will not be much of help, as they will disappear once significant quantities of organic farm products are made available. The demand for organic products is high in the advanced countries of the west like USA, European Union and Japan. It is reported that the US consumers are ready to pay a premium price of 60 to 100 per cent for the organic products. The upper classes in India are also following this trend as elsewhere. The market survey done by the International Trade Centre during 2000 indicates that the demand for organic products is growing rapidly in many of the world markets while the supply is unable to match it.

India is known in the world organic market as a tea supplier and there is a good potential to export coffee, vegetables, sugar, herbs, spices and vanilla. In spite of the several initiatives to produce and export organic produces from the country, the aggregate production for export came to only about 14000 tonnes. This also includes the production of organic spices in about 1000 ha under certification. Some export houses like Good Value Marketing Ltd and Burmah Trading Corporation are also engaged in exporting of organic fruits, vegetables and coffee from India. The country could export almost 85 per cent of the production indicating that demand is not a constraint in the international markets for organic products.

Hybrid seeds are designed to respond to fertilizers and chemicals. The seed, fertilizer and pesticide industry as also the importers of these inputs to the country have a stake in the conventional farming. Their opposition to organic farming stems from these interests. The need for fixing standards and quality parameters for bio-fertilizers and biomanures has arisen with the increasing popularity of organic farming in the country. There are a very large number of brands of organic manures, claiming the high levels of natural nutrients and essential elements. But most farmers are not aware of the pitfalls of using the commercially available biomanure products. While the concept of organic farming itself lays great stress on the manures produced on the farm and the farmers' household, many of the branded products available in the market may not be really organic. lements of chemicals slipping into the manures through faulty production methods could make the product not certifiable as organic. The process of composting which is a major activity to be carefully done is achieved usually by one of the two methods, vermi-composting or microbe composting.

While the former is ideal for segregated waste material without foreign matter, microbe composting is suitable for large scale management of solid wastes, especially in cities and metres. Even though the farmers are using manure produced by different methods, proper parameters for biomanure are yet to be finalized. Most farmers are still unaware of the difference between biomanure and bio-fertilizer, it is point out. While biomanure contains organic matter, which improves the soil quality, bio-fertilizers are nutritional additives separated from the organic material, which could be added to the soil, much like taking vitamin pills. Bio-fertilizers do nothing to enhance soil quality while the loss of soil quality has been the major problem faced by farmers these days.

An understanding of the real costs of erosion of soil and human health, the loss of welfare of both humans and other living things and the computation of these costs are necessary to evaluate the benefits of organic farming. These costs will have to be integrated to a plan for the implementation of organic agriculture. A recent study shows the inappropriateness of the cost and return accounting methods adopted to find out the economics of the organic farming. An economic evaluation of the bad effects of inorganic agriculture and their internalization through environmental taxes is proposed for a market based approach to promote organic farming in India.

Political Interventions

Agriculture in India is subject to political interventions with the objectives of dispensing favours for electoral benefits. Subsidies and other supports from both the Central and state governments, government controlled prices of inputs like chemical fertilizers, the public sector units' dominant role in the production of fertilizers, government support/floor prices for many agricultural products, supply of inputs like power and water either free of cost or at a subsidized rate, etc. are the tools often used to achieve political objectives. Any movement for the promotion of organic farming in India will have to counter opposition from the sections who benefit from such policies in the conventional farming system. The political system in a democracy like India is likely to evade the formulation of policies, which affect the interests of the voting blocks unless there are more powerful counter forces demanding changes. In the absence of alternative employment opportunities and other considerations, the organized workforce particularly in the public sector fertilizer, pesticide and seed industries is also likely to oppose moves on the part of the government to promote organic farming on a large scale.

AGRICULTURAL BENEFITS OF ORGANIC FARMING

Organic agricultural practices are based on a maximum harmonious relationship with nature aiming at the non-destruction of the environment. The developed nations of the world are concerned about the spreading contamination of poisonous chemicals in food, feed, fodder and fibre. Naturally, organic farming system is looked upon as one of the means to remedy these maladies there. However, the major problem in India is the poor productivity of our soils because of the low level content of the organic matter.

The efficiency of the organic inputs in the promotion of productivity depends on the organic contents of the soil. There were many resemblances of organic farming principles in the traditional agriculture of India. But the former gives a more open and verifiable scientific foundation than the latter.

A study conducted in USA on the nutritional values of both organic and conventional foods found that consumption of the former is healthier. Apples, pears, potatoes, corn, wheat and baby foods were analyzed to find out 'bad' elements such as aluminum, cadimum, lead and mercury and also 'good' elements like boron, calcium, iron, magnesium sellenium and zinc. The organic food, in general, had more than 20 per cent less of the bad elements and about 100 per cent more of the good elements.

Increased Crop Productivity and Income

Field trials of organic cotton at Nagpur revealed that during the conversion period, cotton yield was low compared to the conventional and integrated crop management. However, the yields of organic cotton started rising from third year. Cotton yields under organic, conventional and the mixed systems were 898, 623 and 710 kg/ha respectively at the end of the fourth year of the cultivation. The yield of soyabean under organic farming was also the highest compared to the other two systems.

The Central Institute for Cotton Research, Nagpur conducted a study of economics of cotton cultivation in Yavatmal district of Maharashta. The cost of cultivation of cotton was lower in the organic farming than in the modern system. The low costs were due to the non-use of fertilizers and chemical insecticides.

As a result of the low yields during the conversion period, the net income from the organic farm was lesser than the conventional farm. But the yield under organic method increased progressively equalling it to that of the conventional system by the sixth year. The input costs were low under organic farming and with a 20 per cent of premium prices of output, the net income increased progressively from fourth year under organic fanning. The appreciation of net income from organic cotton cultivation by the sixth year was 80 per cent over the conventional crop. Results reported from 1050 field demonstration cum trials under the National Project on Development and Use of Biofertilizers in different parts of the country show* an increase of 4 per cent in yield in plantation crops, 7 per cent in fruit crops, 9 per cent in wheat and sugarcane, 10 per cent in millet and vegetable, 11 per cent in fibre, condiments and spice crops, 14 per cent in oilseeds and flowers and 15 per cent in tobacco.

A study of 100 farmers in Himachal Pradesh during a period of 3 years found that the total cost of production of maize and wheat was lower under organic farming and the net income was 2 to 3 times higher. Both productivity and premium prices contributed to the increased profitability. Another study of 100 farmers of organic and conventional methods in five districts of Karnataka indicated that the cost of organic farming was lower by 80 per cent than that of the conventional one.

Low Incidence of Pests

The study of the effectiveness of organic cotton cultivation on pests at the farm of Central Institute for Cotton Research, Nagpur revealed that the mean monthly counts of eggs, larva and adults of American BoUworm were far lesser under organic farming than under the conventional method. Bio-control methods like the neem based pesticides to Ti-ichoderma are available in the country. Indigenous technological products such as Panchagavya which was experimented at the University of Agricultural Sciences, Bangalore found to control effectively wilt disease in tomato.

Employment Opportunities

Organic farming requires more labour input than the conventional farming system. Thus, India which has a very large amount of labour unemployment and under employment will find organic farming an attraction. Moreover, the problem of periodical unemployment will also get mitigated because of the diversification of the crops with their different planting and harvesting schedules resulting in the requirement of a relatively high labour input.

Indirect Benefits

Several indirect benefits from organic farming are available to both the farmers and consumers. While the consumers get healthy foods with better palatability and taste and nutritive values, the farmers are indirectly benefited from healthy soils and farm production environment. Eco-tourism is increasingly becoming popular and organic farms have turned into such favourite spots in countries like Italy. Protection of the ecosystem, flora, fauna and increased biodiversity and the resulting benefits to all human and living things are great advantages of organic farming which are yet to be properly accounted for.

PROGRESS OF ORGANIC FARMING IN INDIA

The first conference of NGOs on organic farming in India was organized by the Association for Propagation of Indigenous Genetic Resources in October 1984 at Wardha. Several other meetings on organic farming were held at different places in the country towards the end of 1980s. Here, mention must be made of the Bordi Conference in Maharashtra, the state which was the focal point for the organic farming movement in India. The Rajasthan College of Agriculture with the support of the state government organized a meeting on organic agriculture in 1992. The United Planters' Association of South India organised two national level conferences on organic farming in 1993 and 1995. ARISE is a major organization in the country engaged in the promotion of organic farming. ARISE was founded in 1995 at a national conference of organic farming held at Auroville. ARISE comprises of a supporting network of regional groups aiming at sustainable environment by protecting bio-diversity and promoting organic agricultural practices. The selection of Auroville for the conference was apt as it housed the Arabindo Ashram and the pioneering work under its auspices on building technology, alternative energy research, wasteland development, afforestation and organic agriculture.

By 1980, three groups of Indians had taken to organic farming. The first one consisted of urban educated technocrats for peripheral interest, which did not last long. Educated farmers consisted of the second group whose farming practices were based on scientific knowledge. The third group practiced organic farming through trial and error. The successful organic farmers in India are those who have access to sufficient natural resources like, water and other organic inputs mostly on their own farms. These farms produce crops like sugarcane, areca, cocoa, coconut, pepper and spices. Many of them have shown that switch over to organic farming do not affect yields and income and more importantly, knowledge/ expertise is available for successful adoption of organic farming in the country.

The International Federation of Organic Agriculture Movements estimates that an area of about 41,000 hectares in India is under organic farming representing about 0.17 per cent of the world organic acreage. It also reveals that the percentage of organic area to the total cultivated area comes to only about 0.03 per cent and the total number of farms comes to about 5,661. But, a comparison of our 41,000 ha to Australia, Argentina, Italy and USA clearly indicates that organic farming in India has to go very far even to catch up with that of the leading nations of the world.

Non Governmental Organizations are spearheading organic farming in India. A report in 2002 indicates that about 14,000 tonnes of organic products have been raised in India. They include tea, coffee, rice, wheat, pulses, fruits, spices and vegetables. India exports organic agricultural produces to European Union, USA, Canada, Saudi Arabia, UAE, Japan, Singapore and Australia, among others.

The International Conference on "Indian Organic Products-Global Markets" at the end of 2002 was the first to be held in India. IFOAM predicts that India and China have great potential to be organic farm produce exporters in the future. An important event in the history of the modern nascent organic farming in India was the unveiling of the National Programme for Organic Production on 8"^ May, 2000 and the subsequent Accreditation and Certification Programme on P' October, 2001. The logo "India Organic" was released on 26'*' July 2002 to support the NPOP.

An important progress towards organic agriculture made by India is the increasing awareness of the ill effects of the modern farming system, which the country adopted about 35 years ago. The threat poised by the conventional food products to the human health and the damage done to the ecology are being viewed seriously. Efforts are made to produce healthy foods and the demand. for them is increasing. The importance of the marketing of the organic products is highlighted for the promotion of organic agriculture. Several individuals and associations have taken to organic farming and organic products are available in the large cities to a very limited extent.

Important Step Towards in Organic Farming

The most important step towards organic farming taken by the government was to draw a regulatory framework. It is true that the initiatives by the government to introduce organic farming by laying down regulations came belatedly as many countries have already done this kind of basic work decades ago. The implementation of NPOP is ensured by the formulation of the National Accreditation Policy and Programme. The regulations make it mandatory that all organic certification bodies should be accredited by an Accreditation Agency. The international certification agencies operating in India even prior to these regulations will also have to get accreditation under the new dispensation.

The regulations lay down the institutional arrangements for implementing the national programme for organic production. The NPOP is administered, monitored and implemented for the benefit of farmers, processors, traders and consumers. It envisages a three tiered organisation under the overall guidance of the Union Government with the Department of Commerce, Ministry of Commerce and Industry as the nodal agency. Policy making and declaration of the standards for organic products, recognition of organic standards of other nations, efforts to get our standards recognized by others and coordination with other arms of the government for the successful management of the organic agriculture are the major functions entrusted to the ministry.

The agencies accredited are the Agricultural and Processed Food Products Export Development Authority, Coffee Board, Tea Board and the Spices Board. The regulations cover exports, imports and the domestic trade of the organic products. But the government regulations are applicable to only the exports. So, an organic farm product can be exported only if it is certified by a certification body accredited for the purpose. The categories of products covered under accreditation are organic crop production, organic animal production, organic processing operations, wild products and forestry.

A national level steering committee is functioning as the apex advisory body for assisting the government to promote organic farming in the country. This body consists of representative's form the Ministries of Agriculture, Food Processing Industries, Forests and Environment, Science and Technology, Rural Development and Commerce.

Organic production requires certification after periodic inspections in order to ensure that all prescribed practices are followed. The inspection and certification are done by the agencies accredited to the Accrediting Agencies dealing with the commodity. Inspection and certification agencies can be government departments, NGOs, trade or consumer or producer organisations. Such agencies should be registered bodies, with managements in position, declaring the persons who shall be held responsible for any miscarriage of certification and having proof of adequate field staff to undertake periodic inspections. The continued accreditation of such bodies is dependent upon their record of fidelity to the principles of organic production. They are authorised to award certificates after due satisfaction that practices conformed to those enunciated by the Accrediting Agency in relation to the item concerned. The charges levied by the certifying agencies are fixed by the Accrediting Agencies.

SEVERAL PROJECTS AND INITIATIVES TO PROMOTE ORGANIC FARMING

Several projects and initiatives to promote organic farming in the country have begun at the behest of individuals and institutions. A project aided by the World Bank to empower the rural communities in the country to grow organic products for exports had come up in 2002. The programme aims at the improvement and promotion of organic production of spices, certification and export of black pepper, white pepper, ginger, turmeric, cardamom, clove, nutmeg and herbals like rosemary, thyme, oregano and parsley. The implementation of the progamme is done by the NGOs, and Idukki and Waynad districts of Kerala, Nilgiri district of Tamil Nadu and Kandhamal district of Orissa are the areas selected for the purpose. Imparting training to both the JNTGOs and the farmers on organic production methods, basic standards required, documentation, inspection and certification is a major objective of the programme. The assistance to NGOs includes among others computer hardware and software especially for market promotion of their produces.

An initiative for the spread of organic farming by various stakeholders in the Indian organic agriculture sector is their coming together to constitute an apex body for providing centralized services and expertise for the increasing number of organic farmers in the country. Initial steps for setting up the Indian Competence Centre for Organic Agriculture has been taken at a meeting held at the National Academy for Agricultural Sciences in New Delhi in 2003 under the joint auspices of INDOCERT, a Kerala based organic certification agency and the Swiss based FiBL. The decision was the outcome of an Indian team's visit to Switzerland in 2003 to study the structure of Swiss organic farming.

The meeting elected a nine member promoter board of directors to facilitate the setting up of ICCOA, which would be registered as a charitable society. The Centre would strive to strengthen and supplement the efforts of the state governments in promoting the development of organic agriculture in the country.

In Haryana, an enterprising farmer who began farming on his 16 acre land in the Sonepat district in 1971 could establish an organic farm on 108 acres, raising vegetables and other crops. It appears that his success is mainly attributable to the efforts made by him to market the products. NGOs functioning in the neighbourhood of Delhi buy the produces like rice, wheat, pulses and vegetables from his farm at a premium price of 30 to 50 per cent. An exporting firm at a premium of 20 to 30 per cent buys his basmathi rice. Almost 70 per cent of his farm production is sold through advance agreements/ contracts. Presently he and his friends are working with the resident associations in Delhi to market their organic products at a premium of about 25 per cent. In Rajasthan, the Morarka Foundation, established in 1995, promotes sustainable agriculture. It has about 10,000 partners producing vermi-compost and the Foundation is said to be the single largest producer of this organic input in Asia. It encourages the production of bio-pesticides and supports procurement of certified organic products. It has set up a joint venture to promote agri-biotechnology parks and a model park of such a nature has come up in Jaipur on 20 acres of land.

The campaign launched by the Foundation in the Sheldiawati region of the state in favour of organic farming has resulted in reduction of cultivation costs and improvement in the quality of the produces. Small and marginal farmers in the districts of Sikar and Jhunjhunu, who had been complaining of degradation of their agricultural land and declining productivity, were benefited. The efforts of the Foundation to popularize the use of vermi-compost in place of the chemical fertilizer by creating awareness and imparting training to the farmers have been successful. The Foundation too has a large facility to produce vermi-compost. Application of vermi-compost reduces irrigation, increases the flavour of the products and results in a decline in the damage to the crops by insects. The Foundation is presently engaged in developing techniques to enrich vermi-compost through micro-organisms to make it suitable for location and crop specific application. The promotion of organic farming is encouraged through natural resource management based on ecology protection and sustainable agricultural methods.

In Kerala, the POABS Organic Estate at Nelliyampathy is engaged in the organic cultivation of several agricultural produces. Research on organic farming methods including manures and pesticides is also undertaken there. Liquid manures, mixtures of slurry, cow urine and some herbal preparations are also made for use in the estate.

Several people have taken to organic farming in the state fully convinced of its beneficial effects on man and nature. A farmer who took organic farming as a mission in the district of Kottayam was an example. His estate grew only green crops, from vegetables to rubber. When the rubber plantation was raised on organic methods, the yields were low during the initial years. On the application of the organic inputs, they increased and after three years the yields were on par with the conventional rubber trees.

An initiative to make the state of Kerala fully organic has begun with the formulation of a draft policy in 2003. A workshop on Organic Sustainability of Kerala - A Global Model was organised jointly by the state agriculture department and the Confederation of Indian Industry to deliberate on the organic farming in Kerala. The government policy will focus on conversion of land, produces and budgetary support. Assistance will also be available for certification and inputs, promoting the local certifying bodies, development of agronomic practices, extension support and training to farmers. Creation of consumer awareness, quality considerations, and emphasis on income rather than on yield will also be part of the policy. A campaign to promote organic farming on the lines of the programme on Literate Kerala is proposed to be launched. Organic farming, it is suggested, can be promoted among the educated youth who are presently averse to conventional agriculture.

The Peerumedu Development Society is a cooperative society in the Idukki district, which promotes organic farming for the last 12 years. Pepper, cardamom, coffee, tea, nutmeg, clove, etc. are the crops raised by about 20000 farmers of the PDS who are encouraged to adopt organic farming. PDS collects the products from its member farmers and exports too. It has also started an organic tea processing factory in district.

A view currently gaining ground in the state is that organic farming has almost become necessary for the Kerala farmer, reeling under the onslaught of the highly sensitive international markets for agriculture products. The close linking of solid waste management with bio-manure production is opening up new avenues in city planning in Kerala. The State has short listed three organizations for conducting solid waste management programmmes. The possibilities of making fuel pellets out of the rejected wastes are also being explored. 'Jaivam' or organic is the catchword now gaining popularity in Kerala. The tide seems to be turning in favour of bio-manure, bio-pest control and bio-disease control in all forms of agriculture. The state controlled Vegetable and Fruits Promotion Council, Keralam is also taking slow steps in promoting the 'bio' trend as against the overriding importance given to the chemical methods of farming.

An Association of Agriculturists Functioning in the State

'Infam' is an association of agriculturists functioning in the state. As a part of its efforts to promote organic farming, it is planning to launch a producers' company for marketing of organic produces with a view to assist the growing trend of organic farming in the state. In the initial stage, the functioning of the company will be confined to the Wayanad district. The company will have about 3200 farmers of the district as members who have decided to switch over to organic mode of cultivation. Steps are also being taken to forge collaboration with the Indian Institute of Rural Management, Anand. The proposed tie up with IRMA will be for technical help and also for marketing of the products. The Infam has conducted about 300 training programmes in Wayanad district, which had motivated the farmers to adopt organic farming methods of cultivation.

The MS Swaminathan Research Foundation, Chennai that has been operating in the area, assists the movement. Wayanad district is on the verge of an environmental calamity due to the excessive use of chemical fertilizers and pesticides that depleted the soil beyond redemption. The top soil is fast losing its fertility and is becoming barren in many areas. About 2100 banana farmers used tonnes of Furidan, an insecticide annually in the district alone. The increased use of chemicals was also posing a health hazard. A survey conducted by the Infam among the school children in Thavingal Panchayath recently revealed that 250 of them were cancer patients. The cancer disease in the region is alarmingly on the rise. Karnataka has finalized the policy on organic farming and thus has become the first state in the country to adopt it. It encourages the farmers to adopt organic farming and gradually give up the use of chemical fertilizers and pesticides. The government has earmarked Rs 20 crores in the state budget of 2004-05 for the purpose.

The organic farming movement in the state has reached a stage in which the creation of markets for the products has become essential. Association for Promotion of Organic Farming at Bangalore consists of 150 voluntary members. It has been promoting organic farming by creating awareness, inspecting and giving organic certification to the farms. Organic farm production in the state is not very significant. However, some NGOs and individuals are buying organic produces although the movement to convert the land to organic farming has not been successful. There is no organised market for organic produces and if not sold out in the right time, they are left with the conventional produce at the same prices. At the same time, while the farmers are interested in the organic cultivation the consumers are willing to pay for such products. But the link between the two simply does not exist. Consumers demand consistent supplies of good quality organic products and the farmers want an assured market for such produces.

APOP is making the farmers aware of the organic farming methods, particularly about the preparation of the soil for the purpose. The principles behind the healthy soil through the use of the on-farm inputs and compost making are explained to them. Purushotthom Rao who owned 10 acres of land in the Shimoga district took up organic farming in 1989. He raised coconut and paddy along with several other crops in place of arecanut which was destroyed due to strong winds. The yield of coconut he obtained was 80 per cent more than that of his conventional farming friends. The cost of paddy cultivation incurred was 80 per cent lesser than his said friends.

An enterprising farmer practicing organic farming could export several products to the European market. His firm sent 5000 kg of black pepper, 1000 kg of nutmeg, 500 kg of mace, 4000 kg of white pepper, 1000 kg of clove, 300 kg of vanilla, 2000 kg of turmeric, 2000 kg of ginger, 5000 kg of white hibiscus, 80000 kg of henna and 500 kg of cardamom. His firm is certified by a European agency as organic since 1994.

The state of Manipur in the north-east> of India has decided to encourage organic agriculture. The north-east region with its unique characteristics and agricultural practices can be a potential area for the introduction of organic farming. Manipur is aiming at the encouragement of the eco-friendly farm practices and pollution free industry in the state. Dependence of the agro-chemicals will be avoided and factories and industries causing harm to plants and humans will be discouraged. The emphasis on protection of environment is in perfect harmony with the identification of the entire north-east as the mega diversity spot containing rich reserves of flora and fauna. Given the geographical location and agro-climatic zones, the state gives importance to horticulture and agro-industries. The National Horticulture Board has identified 2.77 lakh hectares for the exclusive development of horticulture in the state. It has been decided to encourage organic farming in this development area. The state has proposed to the setting up of an exclusive export zone for organic products in collaboration with the Agriculture Processed Food Export Development Authority. Tamil Nadu plans to encourage organic farming in horticulture and plantation crops to increase the income of the farmers of the state. The state government also wants to promote organic cultivation of fruits, vegetables and tea. The Tamil Nadu Agricultural University has established a model organic farm on a 2.5 hectare area in the campus.

The National Bank for Agriculture and Rural Development has decided to promote organic cultivation of horticultural crops in the mango producing areas of southern Tamil Nadu districts. Theni and Dindigul are major mango growing areas and the fact that the state government has decided to promote an agri export zone for mangoes in these districts will quicken the efforts for organic farming. The aim is to take the benefits of the premium prices for organic products in the developed countries through exports of mangoes. NABARD has funded a farmer in Theni district to the tune of Rs 3.5 crores for organic mango cultivation on a 252 acre area. Identification of land for organic cultivation was a difficult process and it took one year for the farmer to do so and another three years to convince the banks. Search for the ideal planting materials took several months and he incurred an annual cost of Rs 46000 for certification.

A seminar organised at Shri Murugappa Chettiar Research Centre at Chennai focused on the need for natural cultivation of food products. Application of farm manure, biological pest control, companion planting, inter cropping, introduction of beneficial microbes and fungies were advocated for natural food farming. The importance of earthworms and the harm caused to them by chemical farming were highlighted at the discussions. In the history of organic farming in Tamil Nadu, the efforts made in 1970s to revive the traditional methods of agriculture at Chennai were praiseworthy. A study held by MCRC in the city revealed that 58 per cent of the people were aware of the benefits of organic farming and 82 per cent were ready to buy organic foods if the prices were competitive.

A few farmers in Puliangudi village in Tirunelveli district have successfully adopted a package of eco-friendly technologies in the paddy cultivation. It is found that these practices show good results on the indigenous rice varieties. The cost of cultivation has substantially been reduced and the farmers receive a premium for the organic rice. The cost of cultivation worked out comes to about Rs 8,750 per hectare. The price of rice obtained was Rs 30 per kg. The cultivation turned out to be rewarding economically besides being environmentally acceptable. Several farmers are coming forward to practice the method of organic farming after witnessing the results. Another farmer of the same village revealed that he got about 9,250 kg of paddy per ha and no plant protection was done after adoption of the new method. His cost of cultivation worked out was about Rs 12,500 per ha.

In Goa, a co-operative institution, the Adarsha Krishi Sahakari Kharedi Vikri Prakriya Sanstha, pioneered organic farming of cashew and coconut crops. The organisation has 1000 members and is led by its Chairman under whose leadership the organic cultivation is launched. The Cooperative Society has been receiving export orders from traders from abroad for supply of organic cashew nuts. The Society has helped more than 450 cashew growers to convert to organic farming by imparting training and other supports. It also has a processing facility for about 100 tonnes of cashew nuts during the season. The price paid for a kg of organic raw nuts is Rs 35.50. The processed nuts are taken to Mumbai and Delhi by the Society for sale.

The state of Madhya Pradesh has gone for organic soya cultivation. Plans are also ready to convert maize, wheat and pulse cultivation to organic on the principle of crop rotation. Organic soyabean is cultivated in the Malwa region of the state in the districts of Khargone and Dhar on an area of 8,700 hectares under contract farming. Farmers who have not used chemical fertilizers and pesticides during the last two years on the land offered for organic farming are selected. This condition was necessary to restore the soil health to make the land suitable for organic cultivation. Certification of the farm has been done by SKAL, an international agency accredited to APEDA. On getting the certificate, the farm can use the logo 'India Organic' on soyabean and its products like soya oil and deoiled cake.

The production is meant for both the national and international markets. It is said that the conventional yellow soyabean fetches Rs 12,000 per tonne in the local market, while the organic soyabean rules at Rs 34,000 per tonne in Delhi. However, a price premium of 50 per cent is the least expected for the organic soya. This will be financially beneficial to the farmers, even though the productivity of conventional soyabean is put at 800-900 kg, which is higher than 500-600 kg of the organic soya. The genetically modified soya crops dominate the international market for soyabean and its products. However, the demand for organic soya is high ensuring a good premium for exports from the country.

IMPORTANT OF ORGANIC FARMING

This is important because organic farming embodies the elements of a sound agriculture — traditional practices that have been proven over time. In fact, a good, convenient, working definition for organic agriculture is *good farming practice without using synthetic chemicals*. This working definition distinguishes organic practice from the general milieu of agriculture that existed in the pre-chemical era, much of which was exploitative and unsustainable. Organic farming was never intended to be a "throwback" or regressive form of agriculture.

A truly significant event in the history of organics took place in 1962, with the publication of Rachel Carson's *Silent Spring*. *Silent Spring* is a strong and dramatic statement about the impact of pesticides on the environment. It was one of the key documents that gave birth to environmental consciousness in the 1960s and 1970s. When environmentalists and others began looking around for an alternative to pesticides and industrial agriculture, organic farming was there. Not only was it an approach that did not use synthetic pesticides, it also had an attractive counter-culture name that grew to signify a philosophy of living as well as a method of farming.

While *Silent Spring* and the environmental movement were not about organic farming *per se*, they brought it to public consciousness on a vast scale. It is not uncommon, in fact, for some writers to suggest that organic agriculture began with Rachel Carson's book. Though this assertion is untrue, the book clearly played a major role in stimulating industry growth and in altering public perceptions. From the mid-1960s onward, organics was increasingly identified with pesticide issues. It became the idealized alternative for providing clean, healthy food and environmental protection.

NOTIONS OF ORGANIC

As organic farming and marketing entered the 1970s, it began to develop as an industry. As a result, a clearer definition was needed to distinguish it and its products from conventional agriculture. This was no straightforward task. Environmental issues and other alternative agriculture philosophies had created diverse notions about what organic agriculture was and what it should be. A particularly problematic image grew unexpectedly from the anti-pesticide movement of the 1960s.

This was the romantic notion that organic simply meant "doing next-to-nothing." In this exploitative approach, not only were pesticides avoided, sound farming practices that built the soil were also largely ignored. The results achieved on such farms were predictable, as yields were low and the quality poor. These approaches became collectively known as organic by neglect and are a far cry from the responsible farming models proposed by Albert Howard and J.I. Rodale.

It is unclear how many farmers actually chose to farm "by neglect" and advertise themselves as organic over the years. However, this extreme representation of organic agriculture was quickly taken up by critics who tried to characterize all of organic agriculture as soil depleting and unproductive. To counter this, current standards for certified organic production require an "organic plan" outlining the use of soil building activities and natural pest management.

There is a further notion that organic farming also describes farm systems based on soil building, but that continue to use some prohibited fertilizers and pesticides in a limited or selective manner. A USDA study of U.S., organic farms made note of many such individuals who readily and sincerely referred to themselves as organic farmers. While these growers were largely conscientious and would, in most instances, fall under the modern umbrella of "sustainable farmers," industry standards evolved to preclude all synthetic pesticides or commercial fertilizers. The approach to farming by this loose-knit group of growers and their supporters has come to be called "eco-farming" or "eco-agriculture" — terms coined by Acres USA editor Charles Walters, Jr.

NOTION OF ORGANIC AGRICULTURE

A further notion of organic agriculture that bears addressing is the persistent image of organic farming as being possible only on a very small scale. This impression has been enhanced by the high visibility of organic market gardens.

These, of course, are small because market gardening — conventional or organic — is usually done on a smaller scale. Also, some organic market garden systems, such as Biointensive Mini-Farming, use highly labour intensive/low capital investment technologies. These have become popular among U.S., gardeners and, more importantly, with those concerned with Third World development, where such systems are especially relevant. Focus on these systems has, unfortunately, distorted the picture of organics as a whole.

Traditionally, organic farms truly have been smaller than conventional operations. This has been due in part to labour requirements. Organic systems are generally more labour intensive. Studies done in the late 1970s by Washington University, for example, found that about 11 per cent more labour was required per unit of production where agronomic crops were concerned.

This difference can be much greater where horticultural crops are involved, and farm size may be limited accordingly. However, technological innovations in organic horticultural production are helping to narrow the gap. Organic systems are also more information intensive, requiring additional management time in planning, pest scouting, and related activities. For this reason, organic management can be better done if a farm is not too large. Essentially, the notion that organic systems are only possible on very small farms is a false one. Both the Washington University and the USDA studies confirmed this. Given the range of acceptable technologies available, organic agriculture can be sized to fit a wide range of farms and enterprises.

LANDMARK RESEARCH

Throughout its early history, organic agriculture was treated with either hostility or apathy by the USDA, land grant universities, and conventional agriculture in general. Fortunately, the atmosphere for discussing and investigating organics has improved considerably. While it did not become boldly evident until the 1990s, the tide actually began turning in the late 1970s and early 1980s.

A number of factors precipitated this change, among them the growth in the organic industry. Serious money demands seri-ous attention. Also critical, from the perspective of the research community especially, were some landmark studies that lend credibility to organic farming as a truly viable option for American agriculture. The first of these "landmarks" was a series of studies done by Washington University.

Funded by the National Science Foun-dation, this research was motivated by the energy crisis of the 1970s and the effect that higher energy prices would have on agriculture in the nation's Cornbelt. When researchers learned that there were commercial farms that were not dependent on the high-energy inputs of conventional farming, the focus quickly shifted to the study of organics. In addition to the documentation of practices, crop yields, attitudes, and the sustainability indices (cited elsewhere in this publication), the researchers made what was certainly the most astounding discovery of all, that commercial organic farms could be competitive with conventional farms in the conventional marketplace. Arriving on the heels of the Washington University work was another study of great significance done by the USDA. In contrast to the Washington University effort, these researchers chose to extend their survey of farmers nationally and over a wide range of enterprises.

The findings of the USDA study, which were fair, largely positive, and encouraging, kicked open the door for future organic research in ways that a non-land grant/non-USDA entity like Washington University could not. The final report – bound with pastel green cover sheets — was a conspicuous object at alternative agriculture conferences and field days throughout the early 1980s

Also of particular note was a symposium on organic farming held in Atlanta, Georgia, in late 1981. The meeting was spon-sored jointly by the American Society of Agronomy, the Crop Science Society of America, and the Soil Science Society of America — traditionally very conservative entities. It brought together not only representatives of the Washington University and USDA teams, but a surprising number of other researchers clearly interested in the same issues of sustainability and finding a glimmer of hope in organic agriculture.

Organic Certification

In 2002, when the USDA adopted the National Organic Standard that spells out what farmers and food processors must (and must not) do to be certified "organic," the organic industry already had a long history of relying on third-party certifiers to ensure the integrity of their products and practices.

Under this system, a state-run or accredited private agency (the third-party) evaluates farmers and processors to see whether they conform to the standards of the National Organic Programme (NOP). Those who do can then market their products as "USDA Certified Organic" and display the official USDA organic seal on their packaging.

In essence, certification is largely about integrity — assuring that the buyer is getting what he or she is paying for. Certified organic production, then, means production by approved organic methods, with additional pains taken to eliminate contamination with prohibited materials and commingling with conventional products. There is a common misconception that certified organic means "pesticide residue-free." Consumers have a right to expect little or no pesticide residue on certified organic crops because none are used in their production. However, ours is a dirty world in which pesticides and their breakdown products are ubiquitous. This is only to be expected in a national farm system where more than 99 per cent of all applied farm chemicals miss the target organism.

The principles, practices, and tools discussed in the remainder of this publication reflect the guidelines recognized by the NOP, though minor details may vary among third-party certifiers. It is important, therefore, that producers understand their certifying agency's standards well and keep in close touch with a representative. Note that these principles and practices also provide a foundation for other sustainable approaches to crop production, perennial or annual.

ORGANIC PRINCIPLES

There are several compelling principles that characterize certified organic farming. They include *biodiversity*, *integration*, *sustainability*, *natural plant nutrition*, *natural pest management*, and *integrity*. Most organic operations will reflect all of these to a greater or lesser degree. Since each farm is a distinct entity, there is a large degree of variation.

BIODIVERSITY

As a general rule, diverse ecosystems in nature have a higher degree of stability than those with only a few species. The same is essentially true for agroecosystems. Farms with a diverse mix of crops have a better chance of supporting beneficial organisms that assist in pollination and pest management. Diversity above ground also suggests diversity in the soil, providing better nutrient cycling, disease suppression, tilth, and nitrogen fixation. Good organic farmers mimic the biodiversity of nature through practices like intercropping, companion planting, establishment of beneficial habitats, and crop rotation (sometimes referred to as companion planting across time). The effort to increase biodiversity works hand-in-hand with enterprise diversity, which is often (but not necessarily) an objective on organic farms.

Diversification and Integration of Enterprises

The drive to build biodiversity in organic systems encourages diversity among enterprises, but not as isolated or independent entities. Good organic operations integrate their various enterprises. A good example can be seen among Midwestern organic crop and livestock operations. The typical Midwestern organic operation ties the needs of crops and livestock together in a practi-cal and elegant way. The forage and grain needs of ruminant livestock make for a diverse mix of crops. Particularly valuable is the inclusion of le-gume forages for ruminant feed. Forage legumes in rotation fix a sustainable supply of nitrogen in the soil that feeds subsequent non-legume crops in rotation. Manure from the livestock enterprises is conserved as a nutrient resource and recycled back to the crop fields. Farms such as these have the additional advantage of greater economic sustainability, as their risks are spread over several livestock and crop enterprises.

Sustainability

In addition to the greater economic sustainability afforded by enterprise diversification, organic farmers are often able to reap market premiums for certified production. However, since many organ-ic enterprises realize somewhat lower marketable yields, this has not always translated into higher profits or greater economic sustainability. As more and more organic growers enter the marketplace, it is likely that premiums will stabilize at modest levels and may vanish for some crops. Organic producers need to look well ahead and be aware of shifting trends.

As alluded to earlier, many U.S., organic farms per-form well on many of the measurable indicators associated with sustainability, such as energy con-sumption and environmental protection. However, sustainability is an ideal, and the best that can be said is that current organic farms are closer to the ideal than most alternatives — certainly closer than comparable conventional farming operations.

CONSERVATION TILLAGE AND ORGANIC FARMING

Organic agriculture is often characterized as addicted to maximum tillage — with growers using every opportunity to lay the land bare with shovel, plow, or rototiller. This image has been magnified through the popularity of small-scale organic systems like the French Intensive and Biointensive Mini Farming models that espouse double- and triple-digging to create deep rooting beds. While appropriate to such intensive systems, this degree of cultivation is not characteristic of organic agriculture in general. It may surprise some to learn that a large number of organic producers are not only interested in conservation tillage, they have adopted it. This will be a surprise because many believe that conservation tillage always requires herbicides.

The interest in conservation tillage among organic producers in the Cornbelt was well documented in the mid-1970s by Washington University researchers. They noted that the vast majority of organic farmers participating in their studies had abandoned the moldboard plow for chisel plows. Plowing with a chisel implement is a form of *mulch tillage*, in which residues are mixed in the upper layers of the soil, and a significant percentage remains on the soil surface to reduce erosion. Furthermore, a notable number of organic farmers had gone further to adopt ridge-tillage, a system with even greater potential to reduce erosion. It was especially interesting to note that the use of these conservation technologies was almost nil among neighbouring conventional farms at the time. Organic growers were actually pioneers of conservation tillage in their communities.

Among the more well-known of these pioneers were Dick and Sharon Thompson of Boone, Iowa. Their experiences with ridge-tillage and sustainable agriculture became the focus of a series of publications titled *Nature's Ag School*. These were published by the Regenerative Agriculture Association — the forerunner to the Rodale Institute. They are now, unfortunately, out of print.

Research continues to open up new possibilities in conservation tillage for organic farms. New strategies for mechanically killing winter cover crops and planting or transplanting into the residue without tillage are being explored by a number of USDA, land-grant, and farmer researchers. Notable among these is the work being done by Abdul-Baki and Teasdale at the USDA in Beltsville, Maryland — transplanting tomato and broccoli crops into mechanically killed hairy vetch and forage soybeans.

There are also the well-publicized efforts of Pennsylvania farmer Steve Groff, whose no-till system centres on the use of a rolling stalk chopper to kill cover crops prior to planting. Systems like Groff's and Abdul-Baki's are of particular interest because close to 100 per cent of crop residue remains on the soil surface – providing all the soil conservation and cultural benefits of a thick organic mulch.

After blind cultivation, subsequent weed control operations in larger-scale systems can make use of advances in tillage equipment such as rolling cultivators, finger weeders, and torsion weeders that allow tilling close to the plant row. Smaller-scale operations often use wheel hoes, stirrup hoes, and other less capital-intensive hardware.

Determining the amount, the timing, and the kind of tillage to be done can be a balancing act for the organic grower, but experience and observation over time lead to proficiency.

There are downsides to tillage, however, and most organic growers are well aware of them. The most obvious of these is the dollar cost; organic farmers are as concerned as their conventional counterparts about costs of production and strive to minimize expensive field operations. There is also a cost to the soil and environment.

Every tillage operation aerates the soil and speeds the decomposition of the organic fraction. While this may provide a boost to the current crop, it can be overdone and "burn up" the humus reserves in the soil. Excessive tillage can also be directly destructive to earthworms and their tunneling, reducing their benefits to the land. There is also the danger of compaction, even when field operations are well timed.

PRACTICE OF MULCHING

Mulching is a practice often used by organic growers. Traditionally, it entails the spreading of large amounts of organic materials — straw, old hay, wood chips, etc. — over otherwise bare soil between and among crop plants. Organic mulches regulate soil moisture and temperature, suppress weeds, and provide organic matter to the soil. Mulching is most appropriate to small, intensive operations with high-value annual or fruit crops.

A few systems of no-till organic gardening have evolved from the concept of deep, *permanent* mulching. Among these are the well-known Ruth Stout method and *Synergistic Agriculture* — a raised bed system developed by Emilia Hazelip, who adapted concepts from *Permaculture* and the ideas of Masanobu Fukuoka. Mark Cain and Michael Crane, co-owners of Dripping Springs Gardens — an intensive market gardening operation in Arkansas — have adapted Emilia's system to their farm with considerable success.

Plastic mulch, as long as it is removed at end of growing or harvest season, is also permitted in certified organic production. Its use allows larger acreage to be brought more easily under herbicide-free management, though there are serious issues to be addressed.

High-Input Organic Agriculture

At the beginning of this publication, organic farming was described as a system that uses a minimum of off-farm inputs. While that describes most of organic agriculture as it is currently practiced in the U.S., certified organic farming can also entail much greater reliance on off-farm inputs.

Intensive annual strawberry and vegetable systems under plasticulture are good examples. In these systems, traditional rotations and soil building practices are usually employed, followed by clean cultivation and the laying of plastic mulch and drip irrigation tape on shaped beds. During the season, large amounts of soluble organic fertilizers — typically fish-based — are fed to the crop through the drip system (*i.e.*, organic fertigation). At the end of the season, all plastics must be removed from the field, and it is returned to more standard organic management.

Ideally, an off-season cover crop will be planted. Such systems are often exceptionally productive and economically attractive, when organic premiums are good. The high cost of soluble organic fertilizer (typically hundreds of dollars/acre), however, plus the marginally higher cost of pest controls, make such systems largely non-competitive in the conventional marketplace. The labeling of such high-input systems as organic presents a paradox for many proponents of organic agriculture.

It is unclear whether these technological advancements reflect the kind of farming most practitioners and supporters of organics think of as truly "organic." To begin with, the research citing environmental and economic benefits has largely been done on low-input organic systems; it is questionable whether similar findings would be made about high-input systems, especially regarding environmental matters. Of particular note, while low-input organic systems are documented as being more resistant to erosion, fields under plastic mulch are reported to be fifteen times more erodible. Traditional organic farms leach minimal amounts of nitrogen into tile or groundwater; the losses from fields loaded with high levels of soluble organic fertilizers is certain to be greater, but how much greater is unknown. The fossil fuel energy involved in plastic manufacture, transportation, and application may or may not be compensated by reductions in tractor fuel use.

Finally, the lowered capital investment required to produce a crop by traditional organic methods makes this form of farming more accessible to resource-poor farmers and entails less risk in years of crop failure or lack of premiums. These factors are less certain in a high-input system. A further consideration is the issue of plastic disposal following removal. At this time, there are few to no options for recycling, and landfills are the fate of plastic mulches at the end of each season.

While it is unwise to rush to judgement regarding high-input organic farming, it is clear that some adaptations will need to be made, if the traditional character and sustainability benefits of organic farming are to be preserved.

Fire

While fire can be used in a number of ways in organic agriculture, the area of greatest interest is *flame* or *thermal weeding*. In its most common application, torches mounted on a tractor toolbar throw a hot flame at the base of mature (*i.e.*, heat-resistant) plants, over the inter-row area, or both. Tractor speed is adjusted so that weeds are not *burned* so much as *seared*. Searing is sufficient to kill most seedling weeds and uses less fuel. Liquid propane (LP) gas is the fuel most commonly used, though alternatives such as alcohol and methane offer the possibility of on-farm sources.

Supplemental Fertilization

In many organic systems, crop rotation, manuring, green manuring, along with enhanced biological activity in the soil, provide an abundant supply of plant-essential minerals annually. This is especially true on naturally deep and rich prairie soils. It is less true on poorer soils and on those that have been heavily exploited through non-sustainable farming practices. To correct mineral deficiencies in organically managed soils, organic growers often apply ground or powdered rock minerals.

The most commonly used rock mineral is high-calcium aglime. Dolomitic limestone, various rock phosphates, gypsum, sulfate of potash-magnesia, and mined potassium sulfate are also common. These are all significant sources of primary (P, K) and/or secondary (Ca, Mg, S) plant nutrients. The savvy organic grower applies significant amounts of these materials only with the guidance of regular soil testing.

Less common are other rock powders and fines that are limited sources for the major nutrients but are rich in micronutrients or have some other soil-improving characteristic. Among these are glauconite (greensand), glacial gravel dust, lava sand, Azomite®, granite meal, and others.

Supplementary nutrients that include nitrogen are often provided in the form of animal or plant products and by-products such as fish emulsion, blood meal, feather meal, bone meal, alfalfa meal, and soybean meal. Most of these products also supply some organic matter, though that is not the primary reason they are used.

Evaluating Tools and Practices

One basis on which to evaluate the tools or practices one chooses for an organic operation is whether or not they contribute to biodiversity — biodiversity being one of the principal characteristics of a sustainable organic agriculture.

Crop rotation, cover cropping, farmscaping, companion planting, and intercropping are outstanding examples of practices that contribute to biodiversity. They therefore contribute to the long-term stability and sustainability of the farm agroecosystem. Composting and manuring likewise contribute to biodiversity, but since the diversity they promote is mostly in soil biota, it is rather less obvious to the casual observer.

On the other end of the spectrum are practices such as tillage, cultivation, thermal weeding, solarization, and plastic mulching. Such tools significantly reduce diversity in the field and tend to move the system in a less sustainable direction. This, however, does not necessarily make these practices bad choices.

Organic farming has often been called *natural* farming, as it tries to mimic the processes of nature in producing crops and livestock. However, the analogy goes only so far, since most agricultural systems are characterized by a struggle between human and nature, each with a clear notion of what plants and animals the land *ought* to support and in what proportions. The farmer is typically in the position of "holding back" the natural succession of plant and animal species through the use of diversity-reducing tools and practices. The ideal is to bring about an agricultural system in which the long-term direction emphasizes diversity and sustainability. Among the best visualizations of this ideal are those emerging from the Permaculture movement.

Balanced Nutrition

Whether or not it has been customary in the past, organic growers are encouraged to have periodic soil testing done on their fields. How the results of a soil test are used, however, can vary considerably among farmers, depending on their personal philosophy and management skill.

While there is certainly a segment of the organic farming community that has no faith in soil audits, most growers use them as a means to monitor progress in building their soils, to identify nutrient deficits, and to guide supplementary fertilization. While there are many ideas about fertilization guidelines, there are two schools of thought that dominate.

The first is commonly known as the *sufficiency* approach or model. Though somewhat oversimplified, the following are among its significant characteristics:

- Annual fertilizer additions of P and K are based in good part on how much the crop is expected to remove from the soil at harvest
- Additional amounts of P and K are recommended based on keeping the soil nutrient reserves at a particular level
- Lime is added to the soil based on pH
- Little to no attention is paid to nutrient balance or to the levels of secondary nutrients Ca and Mg.

The second approach is referred to variously as *cation nutrient balancing*, the *Albrecht system*, the *CEC*, or the *base saturation* approach. It differs from the sufficiency approach in that fertilizer and lime recommendations are made based on an idealized ratio of nutrients in the soil and its capacity to hold those nutrients against leaching. Cation nutrient balancing is more popular among practitioners of organic farming and sustainable agriculture in general than it is among conventional growers. However, there is no universal agreement on which approach is most appropriate for organic management.

A wide range of other products —humates, humic acids, enzymes, catalyst waters, bioactivators, surfactants, to name a few — are also acceptable in organic crop production. However, the tradeoff between out-of-pocket cost and efficacy of such materials is often challenged by conventional and organic growers alike. Organic growers are encouraged to experiment, but to do so in a manner that allows the actual results to be measured.

Biorational Pesticides

While, in principle, any pesticide use is discouraged in organic systems, a rather wide range of biorational pesticides is permitted. The frequency of pesticide use varies considerably with crop and location. For example, there is virtually no use of pesticides on organic row crops in the Cornbelt. By contrast, organic tree fruits in the Midsouth routinely receive heavy applications of several fungicides and insecticides allowable in organic production.

The pesticides permitted in organic farming fall predominantly into several classes.

- *Minerals:* These include sulfur, copper, diatomaceous earth, and clay-based materials like Surround®.

- *Botanicals:* Botanicals include common commercial materials such as rotenone, neem, and pyrethrum. Less common botanicals include quassia, equisetum, and ryania. Tobacco products like Black-Leaf 40® and strichnine are also botanicals but are prohibited in organic production due to their high toxicity.
- *Soaps:* A number of commercial soap-based products are effective as insecticides, herbicides, fungicides, and algicides. *Detergent*-based products are *not* allowed for crop use in organic production.
- *Pheromones:* Pheromones can be used as a means to confuse and disrupt pests during their mating cycles, or to draw them into traps.
- *Biologicals:* One of the fastest-growing areas in pesticide development, biopesticides present some of the greatest hope for organic control of highly destructive pests. Among the most well-known biopesticides are the *Bacillus thuringiensis* (Bt) formulations for control of lepidopterous pests and Colorado potato beetle.

ECONOMICS OF ORGANIC FARMING

The replacement of external inputs by farm-derived resources normally leads to a reduction in variable input costs under organic management. Expenditure on fertilizers and sprays is substantially lower than in conventional systems in almost all the cases. In a few cases, higher input costs due to the purchase of compost and other organic manure have been reported. Studies have shown that the common organic agricultural combination of lower input costs and favourable price premiums can offset reduced yields and make organic farms equally and often more profitable than conventional farms. Studies that did not include organic price premiums have given mixed results on profitability.

Studies from Europe and Canada show labour costs in organic agriculture average 40–50 per cent higher where the wage rates are generally higher30. Gross margins, the difference between farm output and variable costs are generally similar or, where there are favourable price premiums, higher in organic agriculture. The economics of organic cotton cultivation over a period of six years indicated that there is a reduction in cost of cultivation and increased gross and net returns compared to conventional cotton cultivation in India.

ORGANIC CERTIFICATION

Organic certification is a certification process for producers of organic food and other organic agricultural products. In general, any business directly involved in food production can be certified, including seed suppliers, farmers, food processors, retailers and restaurants.

Requirements vary from country to country, and generally involve a set of production standards for growing, storage, processing, packaging and shipping that include:

- Avoidance of most synthetic chemical inputs, genetically modified organisms, irradiation, and the use of sewage sludge;
- Use of farmland that has been free from synthetic chemicals for a number of years;
- Keeping detailed written production and sales records;
- Maintaining strict physical separation of organic products from non-certified products;
- Undergoing periodic on-site inspections.

In some countries, certification is overseen by the government, and commercial use of the term organic is legally restricted. Certified organic producers are also subject to the same agricultural, food safety and other government regulations that apply to non-certified producers.

PURPOSE OF CERTIFICATION

Organic certification addresses a growing worldwide demand for organic food. It is intended to assure quality and prevent fraud, and to promote commerce. While such certification was not necessary in the early days of the organic movement, when small farmers would sell their produce directly at farmers' markets, as organics have grown in popularity, more and more consumers are purchasing organic food through traditional channels, such as supermarkets. As such, consumers must rely on third-party regulatory certification. For organic producers, certification identifies suppliers of products approved for use in certified operations. For consumers, "certified organic" serves as a product assurance, similar to "low fat", "100 per cent whole wheat", or "no artificial preservatives".

Certification is essentially aimed at regulating and facilitating the sale of organic products to consumers. Individual certification bodies have their own service marks, which can act as branding to consumers—a certifier may promote the high consumer recognition value of its logo as a marketing advantage to producers. Most UK certification bodies operate organic standards that meet the UK government's minimum requirements. Some certification bodies, such as the Soil Association, certify to higher standards.

THE CERTIFICATION PROCESS

To certify a farm, the farmer is typically required to engage in a number of new activities, in addition to normal farming operations:

- Study the organic standards, which cover in specific detail what is and is not allowed for every aspect of farming, including storage, transport and sale.
- *Compliance*: Farm facilities and production methods must comply with the standards, which may involve modifying facilities, sourcing and changing suppliers, etc.

- *Documentation*: Extensive paperwork is required, detailing farm history and current set-up, and usually including results of soil and water tests.
- *Planning*: A written annual production plan must be submitted, detailing everything from seed to sale: seed sources, field and crop locations, fertilization and pest control activities, harvest methods, storage locations, etc.
- *Inspection*: Annual on-farm inspections are required, with a physical tour, examination of records, and an oral interview.
- *Fee*: An annual inspection/certification fee.
- *Record-keeping*: Written, day-to-day farming and marketing records, covering all activities, must be available for inspection at any time.

In addition, short-notice or surprise inspections can be made, and specific tests may be requested.

For first-time farm certification, the soil must meet basic requirements of being free from use of prohibited substances for a number of years. A conventional farm must adhere to organic standards for this period, often, two to three years. This is known as being in transition. Transitional crops are not considered fully organic.

Certification for operations other than farms is similar. The focus is on ingredients and other inputs, and processing and handling conditions. A transport company would be required to detail the use and maintenance of its vehicles, storage facilities, containers, and so forth. A restaurant would have its premises inspected and its suppliers verified as certified organic.

CERTIFICATION AND PRODUCT LABELING

In some countries, organic standards are formulated and overseen by the government. The United States, the European Union and Japan have comprehensive organic legislation, and the term "organic" may be used only by certified producers. Being able to put the word "organic" on a food product is a valuable marketing advantage in today's consumer market, but does not guarantee the product is legitimately organic. Certification is intended to protect consumers from misuse of the term, and make buying organics easy. However, the organic labeling made possible by certification itself usually requires explanation. In countries without organic laws, government guidelines may or may not exist, while certification is handled by non-profit organizations and private companies.

Internationally, equivalency negotiations are underway, and some agreements are already in place, to harmonize certification between countries, facilitating international trade. There are also international certification bodies, including members of the International Federation of Organic Agriculture Movements, the Organic Crop Improvement Association and Ecocert. Where formal agreements do not exist between countries, organic product for export is often certified by agencies from the importing countries, who may establish permanent foreign offices for this purpose.

North America

In the US, federal organic legislation defines three levels of organics. Products made entirely with certified organic ingredients and methods can be labeled "100 per cent organic". Products with at least 95 per cent organic ingredients can use the word "organic". Both of these categories may also display the USDA organic seal. A third category, containing a minimum of 70 per cent organic ingredients, can be labeled "made with organic ingredients". In addition, products may also display the logo of the certification body that approved them. Products made with less than 70 per cent organic ingredients can not advertise this information to consumers and can only mention this fact in the product's ingredient statement. Similar percentages and labels apply in the EU.

In the US, the National Organic Programme was enacted as federal legislation in Oct. 2002. It restricts the use of the term "organic" to certified organic producers. Certification is handled by state, non-profit and private agencies that have been approved by the US Department of Agriculture. One of the first organizations to carry out organic certification in North America was the California Certified Organic Farmers, founded in 1973. In Canada, certification was implemented at the federal level on June 30, 2009. Mandatory certification is required for agricultural products represented as organic in import, export and inter-provincial trade, or that bear the federal organic logo. In Quebec, provincial legislation provides government oversight of organic certification within the province, through the Quebec Accreditation Board.

Europe

EU countries acquired comprehensive organic legislation with the implementation of the EU-Eco-regulation 1992. Supervision of certification bodies is handled on the national level. In March 2002 the European Commission issued a europeanwide label for organic food however for most of the countries it was not able to replace existing national product labels. It was relaunched in 2009 with a design competition for a new logo to be used throughout the EU from July 2010. The final logo will be chosen through a public vote online, which is open until the 31st of January 2010.

In the United Kingdom, organic certification is handled by a number of organizations, regulated by The Department for Environment, Food and Rural Affairs, of which the largest are the Soil Association and Organic Farmers and Growers.

In Sweden, organic certification is handled by the organisation with members such as farmers, processors, trade and also consumer, environmental and animal welfare interests. In Ireland, organic certification is available from the Irish Organic Farmers and Growers Association In Greece, organic certification is available from eight organizations approved by EU. The major of them are BIOHELLAS and the DIO.

In France, organic certification was introduced in 1985. It has established a green-white logo of "AB - agriculture biologique". The certification for the AB label fulfills the EU regulations for organic food. The certification process is overseen a public institute established in November 2001. The actual certification authorities include a number of different institutes like Aclave, Agrocert, Ecocert SA, Qualité France SA, Ulase, SGS ICS. In Belgium, a similar process as in France is being used where certification of the "Biogarantie" label is overseen by the public "association sans but lucratif" administration. This administration does exist since 27. June 1921 and it was reformed on 2. May 2002 to take over the new responsibilities of the label certification.

In Germany the national label was introduced in September 2001 following in the footsteps of the political campaign of "Agrarwende" led by minister Renate Künast of the Greens party. This campaign was started after the mad-cow disease epidemic in 2000. The effects on farming are still challenged by other political parties. The national "Bio"-label in its hexagon green-black-white shape has gained wide popularity - in 2007 there were 2431 companies having certified 41708 products. The popularity of the label is extending to neighbouring countries like Austria, Switzerland and France.

In the German-speaking countries there have been older non-government organizations that had issued labels for organic food long before the advent of the EU organic food regulations. Their labels are still used widely as they significantly exceed the requirements of the EU regulations. An organic food label like "demeter" from Demeter International has been in use since 1928 and this label is still regarded as providing the highest standards for organic food in the world. Other active NGOs include Biokreis, Bioland, Biopark, Ecoland, Ecovin, Gäa e.V. and Naturland. The farmland converted to produce certified organic food has seen a signification evolution in the EU15 rising from 1.8 per cent in 1998 to 4.1 per cent in 2005. For the current EU25 however the statistics do report an overall percentage of just 1.5 per cent as of 2005. Other than the percentage of farmland the statistics show a much larger percentage of organic food in terms of turnover reaching 10 per cent in France and 14 per cent in Germany. Vegetables, fruits, milk and eggs have reached a percentage of 21 per cent to be certified organic food in France. Non-EU countries have widely adopted the European certification regulations for organic food, to increase export chances to EU countries.

In Other Countries

In Japan, the Japanese Agricultural Standard (JAS) was fully implemented as law in April, 2001. This was revised in November 2005 and all JAS certifiers were required to be re-accredited by the Ministry of Agriculture.

In Australia, the Australian Quarantine and Inspection Service is the controlling body for organic certification because there are no domestic standards for organic produce within Australia. Currently the government only becomes involved with organic certification at export, meaning AQIS is the default certification agency.

Although there is no system for monitoring the labeling of organic produce sold within Australia, this primarily affects the retail public. Commercial buyers for whom this is an issue have simply taken the export system as a *de facto* standard and are willing to pay premium prices for produce from growers certified under the National plans. As of 2006, there are seven AQIS-approved certifying organisations authorised to issue Organic Produce Certificates, and in 2004 there were 2345 certified operators. The largest importer of Australia's organic produce is Japan, followed by the UK, France, and New Zealand. The largest certifier of organic products is Australian Certified Organic, which is a subsidiary of Biological Farmers Australia, the largest organic farmers' collective in the country.

In India, APEDA regulates the certification of organic products as per National Standards for Organic Production. "The NPOP standards for production and accreditation system have been recognized by European Commission and Switzerland as equivalent to their country standards. Similarly, USDA has recognized NPOP conformity assessment procedures of accreditation as equivalent to that of US. With these recognitions, Indian organic products duly certified by the accredited certification bodies of India are accepted by the importing countries." In China, the China Green Food Development Centre awards two Standards: A and AA; while the former standard does permit some use of synthetic agricultural chemicals, the latter is more stringent.

CERTIFICATION ISSUES

Organic certification is not without its critics. Some of the staunchest opponents of chemical-based farming and factory farming practices also oppose formal certification. They see it as a way to drive independent organic farmers out of business, and to undermine the quality of organic food. Other organizations such as the Organic Trade Association work within the organic community to foster awareness of legislative and other related issues, and enable the influence and participation of organic proponents.

Obstacle to Small Independents

Originally, in the 1960s through the 1980s, the organic food industry comprised mainly small, independent farmers, selling locally. Organic "certification" was a matter of trust, based on a direct relationship between farmer and consumer. Critics view regulatory certification as a potential barrier to entry for small producers, by burdening them with increased costs, paperwork, and bureaucracy.

Manipulation of Regulations

Critics of formal certification also fear an erosion of organic standards. Provided with a legal framework within which to operate, lobbyists can push for amendments and exceptions favourable to large-scale production, resulting in "legally organic" products produced in ways similar to current conventional food. Combined with the fact that organic products are now sold predominantly through high volume distribution channels such as supermarkets, the concern is that the market is evolving to favour the biggest producers, and this could result in the small organic farmer being squeezed out.

Manipulation of certification regulations as a way to mislead or outright dupe the public is a very real concern. Some examples are creating exceptions and creative interpretation of standards to meet the letter, but not the intention, of particular rules.

A complaint filed with the USDA in February 2004 against Bayliss Ranch, a food ingredient producer and its certifying agent, charged that tap water had been certified organic, and advertised for use in a variety of water-based body care and food products, in order to label them "organic" under US law. Steam-distilled plant extracts, consisting mainly of tap water introduced during the distilling process, were certified organic, and promoted as an organic base that could then be used in a claim of organic content. The case was dismissed by the USDA, as the products had been actually used only in personal care products, over which the department at the time extended no labeling control. The company subsequently adjusted its marketing by removing reference to use of the extracts in food products. Several months later, the USDA extended its organic labeling to personal care products; this complaint has not been refiled.

In December 2005, the 2006 agricultural appropriations bill was passed with a rider allowing 38 synthetic ingredients to be used in organic foods. Among the ingredients are food colourings, starches, sausage and hot-dog casings, hops, fish oil, chipotle chili pepper, and gelatin. This allowed Anheuser-Busch in 2007 to have its Wild Hop Lager certified organic "even though uses hops grown with chemical fertilizers and sprayed with pesticides."

Misrepresentation of the Term Organic

The word organic is central to the certification process, and this is also questioned by some. Where organic laws exist, producers cannot use the term legally without certification. To bypass this legal requirement for certification, various alternative certification approaches, using currently undefined terms like "authentic" and "natural" instead of "organic", are emerging. In the US, motivated by the cost and legal requirements of certification, the private farmer-to-farmer association, Certified Naturally Grown, offers a "non-profit alternative eco-labelling programme for small farms that grow using USDA Organic methods but are not a part of the USDA Certified Organic programme." In the

UK, the interests of smaller-scale growers who use 'natural' growing methods are represented by the Wholesome Food Association, which issues a symbol based largely on trust and peer-to-peer inspection. A related concern holds that certification is replacing consumer education, and this goes against the essential, holistic nature of organic farming. By reducing complex issues and regulations to a simple, convenient certified organic label, consumers may more easily ignore the principles and practices behind organics, leaving the definition of organic farming and organic food open to manipulation.

NUTRIENT MANAGEMENT IN ORGANIC FARMING

Organic farming is often understood as a form of agriculture with use of only organic inputs for the supply of nutrients and management of pests and diseases. In fact, it is a specialized form of diversified agriculture, wherein problems of farming are managed using local resources alone. The term organic does not explicitly mean the type of inputs used; rather it refers to the concept of farm as an organism. Often, organic agriculture has been criticized on the grounds that with organic inputs alone, farm productivity and profitability might not be improved because the availability of organic sources is highly restricted7. True, organic resources availability is limited; but under conditions of soil constraints and climate beggaries, organic inputs use has proved more profitable compared to agrochemicals.

Organic farming systems rely on the management of soil organic matter to enhance the chemical, biological and physical properties of the soil. One of the basic principles of soil fertility management in organic systems is that plant nutrition depends on 'biologically-derived nutrients' instead of using readily soluble forms of nutrients; less available forms of nutrients such as those in bulky organic materials are used. This requires release of nutrients to the plant via the activity of soil microbes and soil animals. Improved soil biological activity is also known to play a key role in suppressing weeds, pests and diseases.

Animal dung, crop residues, green manure, biofertilizers and bio-solids from agro-industries and food processing wastes are some of the potential sources of nutrients of organic farming. While animal dung has competitive uses as fuel, it is extensively used in the form of farmyard manure. Development of several compost production technologies like vermicomposting, phosphocomposting, N-enriched phosphocomposting, etc. improves the quality of composts through enrichment with nutrient-bearing minerals and other additives. These manures have the capacity to fulfill nutrient demand of crops adequately and promote the activity of beneficial macro- and micro-flora in the soil.

There are several doubts in the minds of not only farmers, but also scientists about whether it is possible to supply the minimum required nutrients to crops through organic sources alone. Even if it is possible, how are we going to mobilize the organic matter? At this juncture, it is neither advisable nor feasible to recommend the switchover from fertilizer use to organic manure under all agro-ecosystems. Presently, only 30 per cent of our total cultivable areas has irrigation facilities where agrochemicals use is higher compared to rain-fed zones. It is here that ingenuity and efforts are required to increase crop productivity and farm production despite recurrence of environmental constraints of drought and water scarcity.

The basic requirement in organic farming is to increase input use efficiency at each step of the farm operations. This is achieved partly through reducing losses and adoption of new technologies for enrichment of nutrient content in manure. Technologies to enrich the nutrient supply potential from manure, including farmyard manure three to four times are being widely used in organic farms. Around 600 to 700 million tonnes (mt) of agricultural waste is available in the country every year, but most of it is not used properly. We must convert our filth into wealth by mobilizing all the biomass in the rural and urban areas into bioenergy to supply required nutrients to our starved soil and fuel to farmers. India produces about 1800 mt of animal dung per annum. Even if two-thirds of the dung is used for biogas generation, it is expected to yield biogas not less than 120 m m3 per day. In addition, the manure produced would be about 440 mt per year, which is equivalent to 2.90 mt N, 2.75 mt P2O5 and 1.89 mt K2O.

Organic farms and food production systems are quite distinct from conventional farms in terms of nutrient management strategies. Organic systems adopt management options with the primary aim to develop whole farms, like a living organism with balanced growth, in both crops and livestock holding. Thus nutrient cycle is closed as far as possible. Only nutrients in the form of food are exported out of the farm. Crop residues burning is prohibited; so also the unscientific storage of animal wastes and its application in the fields. It is, therefore, considered more environment friendly and sustainable than the conventional system. Farm conversion from high-input, chemical-based system to organic system is designed after undertaking a constraint analysis for the farm with the primary aim to take advantage of local conditions and their interactions with farm activities, climate, soil and environment, so as to achieve closed nutrient cycles with less dependence on off-farm inputs. This implies that the only nutrients leaving the farm unit are those for human consumption.

Crop rotations and varieties are selected to suit local conditions having the potential to sufficiently balance the nitrogen demand of crops. Requirements for phosphorus, sulphur and micronutrients are met with local, preferably renewable resources. Organic agriculture is, therefore, often termed as knowledge-based rather than input-based agriculture. Furthermore, organic farms aim to optimize the crop productivity under a given set of farm conditions. This is in contrast to concept of yield maximization through the intensive use of agrochemicals, irrigation water and other off-farm inputs. There are ample evidences to show that agrochemical-based, high-input agriculture is not sustainable for long periods due to gradual decline in factor productivity, with adverse impact on soil health and quality.

ORGANIC PRODUCTS TRADED IN THE INTERNATIONAL MARKET

The important organic products traded in the international market are dried fruits and nuts, processed fruits and vegetables, cocoa, spices, herbs, oil crops and derived products, sweeteners, dried leguminous products, meat, dairy products, alcoholic beverages, processed foods and fruit preparations. Cotton, cut flowers, animals and pot plants are major non-food products in the world markets.

Organic farming is bound to grow around the world as many countries are developing their own standards and regulations. The US and the European Union have very comprehensive National Organic Programmes and the early nineties have seen organic farming regulations in Japan, Canada and Australia. New Zealand, Israel and Brazil have adopted the organic standards equivalent to those of USA and the European Union. China, Thailand, South Korea, Philippines, Turkey and Mexico have established certifying agencies. India too has adopted the National Programme for Organic Production with national standards. There were about 35,000 certified organic farms in 17 countries during 1998 covering an area of more than a million hectares. A survey held recently revealed that 6-7 per cent farms in USA and Germany are certified organic farms. Many organic inspection and certifying agencies like Crop Improvement Association, IFOAM, Ecocert, BCS, SKAL. KRAV, IMO, FVO, QAI, Oregon, Tilth and NASAA have strong international presence and have even developed the ability to assist marketing of organic products world wide.

Certification of products is the foundation on which organic agriculture rests. Its role in production and marketing of organic foods becomes a significant corner stone in the creation of a confident consumer community. The need for a trustworthy and healthy organic market is important as the consumers are becoming increasingly separated from the food production process. The void is filled in by a plethora of processing, distribution and marketing intermediaries. The consumers should possess knowledge and confidence that the products they buy are truly organic. Such environment can be created by an organised system of inspection and certification supported by law and regulations of the state. Organic products enjoy trust based on the label or logo approved by the certification process.

The aim of organic certification is, therefore, to:

- Provide an identifiable label (a logo or a symbol); and
- Give an assurance to the consumers that the product is truly organic.

It is widely recognised the world over that the certification of organic products should be based on the following principles:

- Organic production and processing standards should be clearly laid down.
- The conformation of production and processing to these stands must be verified.
- Organic labels should be permitted only to those produces, which are found conforming to the set standards.

Thus, a label on an organic product conveys that the manufacturer has a license for organic production, an independent agency has inspected the production/processing practices followed by the producer, and compliance of the set standards is made. Such an assurance becomes crucial in the generation of consumer confidence in the organic products, particularly as they are costlier than the conventional ones.

The marketing agents of organic products also need such an assurance to maintain fair trade practices essential to the growth and sustainability of the demand for such products.

GROWTH OF ORGANIC FARMING

Organic farming has spread to about 100 countries around the world. An estimate in 2004 puts about an area of 24 million hectares under oi'ganic farming worldwide. Australia with its 10.5 million hectares leads the countries. However, much of this area is pastoral land for grazing. Argentina with 3.19 million hectares, Italy with 1.83 million hectares and USA with 0.95 million hectares follow suit. The importance of organic farming is growing in many countries. Austria and Switzerland have about 10 per cent of their food system under organic agriculture. The annual growth of organic farming is estimated to be about 20 per cent in USA, France, Japan and Singapore.

ORGANIC FARMING IN USA

Organic farming did spread fast in USA during the 1990s. The growth was in response to the rising demand for organic products both in the national and international markets. The US agriculture farms view organic farming as a method to reduce input costs and use of non-renewable resources, which increases accessibility to the high value markets and assures premium prices for the produces and above all increased farm income. The US organic farming system uses eco-friendly agriculture practices in cultural and pest management and avoids the application of chemicals. Antibiotics and hormones are not used in livestock production.

The growth of consumer demand for organic products has also resulted in the specialization in growing, processing and marketing of these products. The implementation of national organic standards in 2002 by the USDA has further boosted the demand for organic products in the country. Because of increasing prices and ineffectiveness of chemicals, many farms in USA do not favour intensive chemical based agi-iculture. A survey of certified organic farms revealed that more than 60 per cent of them grew vegetables and they had livestock for manure to use in the farm. The farmers had identified three problem areas, which required study.

They were:

- The need for increasing demand for organic products,
- The influence of cultural practices on quality and nutritional value of the products; and
- The relationship between plant nutrition and resistance to pests.

In another study of consumer demand for organic foods, it was found that about 30 per cent of shoppers used some organic food in a year, while 10 per cent used it once a week. The value of sales of organic products in US was $1.5 billion in 1992 and $8 billion in 2000. Many methods to promote sales and consumption of organic food have been introduced. The US Senate approved a bill to make a part of the Federal School Lunch Programme organic. Organic food shows are held annually to give awards for the best organic foods and drinks. The Federal government encourages the planting of native plants for Municipal planting programmes. The government has initiated polices to reduce the application of pesticides and labelling food to indicate the amount of pesticides used to produce the commodity is compulsory. There are about 17 organic certification agencies in US.

The certification and labelling laws are very rigid in United States. Imports of organic products must conform to the standards laid down by the USDA and naturally, the ability of many developing countries to comply with these requirements is limited. A grower or a dealer in need of certification has to lodge a plan to an accredited agency of the USDA. There are both private and state certification programmes. The plan must contain all details in relation to the farming or handling methods and the materials those will have to be used therein. The anticipated improvements and future aims should also be a part of the plan. The farmers or dealers of wild crops should also submit a plan mentioning that the harvesting techniques to be employed will not cause any damage both to the produce and the environment. All records related to the management practices and the materials utilized in organic production should be maintained for five years. Certification will be approved only if the crops are grown on the land free of prohibited substances for three years prior to harvest. The period of conversion is three years between the conventional and organic farming and the crops grown during this period cannot be labelled organic or any other label indicating that they are under conversion.

The Organic Foods Production Act contains a wide and detailed coverage of organic farming methods and materials along with the soil fertility, application of manure, crop rotation and composting. The National Organic Standards Board's recommended ingredients include crop residues, crop wastes from food processing operations, animal manure, yard wastes from private and municipal sources, or other vegetable by-products. Municipal solid waste compost and sewage sludge composts and the use of prohibited material as composts/ ingredients are banned. There is a National List of acceptable and banned materials which includes pest control treatments and other agricultural inputs like fertilizers and seed treatments. An evaluation of all agricultural inputs has been made for their long term effect on the environment irrespective of the fact that they are synthetic or natural.

The exporters of other countries to the US market will have to face these very tough specifications, particularly if their national standards are not equivalent to those of the USDA.

ORGANIC FARMING IN ITALY

The European Union of 25 countries had 5.5 million hectares under organic farming at the end of 2002 managed by 1,60,000 organic farms. This constitutes about 3.4 per cent of the agricultural area and 1.7 per cent of farms in EU. It is estimated that the growth in the area was 9 per cent over the last year. The increases came mainly from France, Spain and the United Kingdom. The European Commission has formulated the 'European Action Plan for Organic Food and Farming' on 10* June 2004. It proposes 21 specific actions to promote organic farming in the EU countries and the Action Plan is with the European Parliament for consideration. As said earlier, the progress achieved by Italy, Austria and Germany was impressive.

The pioneers of organic farming in Italy had contributed immensely to the spread of organic agriculture, which is acclaimed to be one of the best in the world. Farmers, medical doctors, food processors, traders and consumers were in the forefront working hard from the 1950s for the cause of organic farming in the country.

Growth of organic farming in Italy has been spectacular during the period of 1990-2000. The organically cultivated area increased from 13,000 ha to more than a million ha which came to about 7.2 per cent of the total cultivated area. The number of farmers was about 13,00 in 1990, which shot up to 47,357 at the end of 2000. At the end of 2002, the area increased to 1,83,000 ha and the number of farms to 56,440. The reasons for such a substantial increase were the food scandals like the mad cow disease leading to the search for safe foods, the constant endeavor for alternative farming methods which were environmentally, technically and economically attractive to the farmers and the liberal subsidies.

The farmers who opted for organic methods prior to 1990 were people of non-agriculture background, people searching for novel life styles, etc. A significant proportion of them consisted of non-Italians like, Germans, Swiss,

Austrians and Americans. They had the unity of purpose in establishing cultural associations in cooperation with like minded people by setting up marketing links with organic farms. However, the decade of 1990-2000 saw that the new entrants into organic farming were Italians and many of them did not have experience of the new methods. But they possessed a good knowledge of the ecology and understood the need for a marketing system. The period also witnessed a change in the geographical distribution of area under organic cultivation. Sardinia became an important region in organic farming with the extension of subsidies to pasture lands used for sheep rearing and milk production. Sardina and Sicily account for about 44 per cent organic area in Italy.

At present 41 per cent of the organic area is under pastures. Cereal crops are raised on 18.2 per cent of the area and the olive cultivation alone has the highest single crop share of 10 per cent. The high growth of organic production led to the growth of the processing units to the extent of 142 per cent during 1997-2000 alone. Presently, there are about 42,000 processing units of which 1,300 are established by the farmers themselves. Organic farming has resulted in the setting up, conversion or expansion of new economic activities in the rural areas of Italy.

The first organic shop had come up in Milan, the industrial and financial capital of Italy, during the 1970s under a consumers' cooperative venture. The most important fact of the market distribution of organic products in Italy is the predominant position of the farmers' or consumers' cooperatives. The distribution was mostly done during the immediate post 1970s by small grocery stores and the major share was that of the direct sales in the open market places and at the farms. After 1990, specialized markets, supermarkets, franchisees, school canteens, etc. became the major channels. Supermarkets are entering the field in large numbers since 1999 with their own labels to convince the consumer of the reliability of the products in the midst of numerous labels. Diary products have a share of 26 per cent of the consumer expenditure on organic products; fruits and vegetables accounts for 13 per cent; breads and biscuits had a share of 12 per cent. A growing number of kindergartens, school and university canteens, restaurants and factory Ccinteens are opting for organic foods in the country. A survey of the motives of the consumers of organic foods revealed that every one was going for organic products to protect the health of his/her family.

The support for organic farming by the European Union by various legislations providing subsidies was instruinental in the growth of organic farming in Italy. The devolution of power to the Regional Government on agriculture policy in 1972 has resulted in different levels of subsidies in different regions/crops. There are nine organic certification bodies in Italy. Some of them were earlier cultural associations promoting organic farming. A single agency named ICEA has a share of 24 per cent of the certified farms coming to about 29 per cent of the area. These nine agencies have about 900 local offices, 1000 staff and technicians for inspecting farms, processing units and storages for giving certifications.

Universities and Public Research Centres are becoming increasingly interested in research and extension activities in relation to organic agriculture. A National School of Organic Farming has been established by the Ministry of Agriculture to impart education.

ORGANIC FARMING IN AUSTRIA

Austria is a leading country in Europe in the adoption of organic farming. At the end of 2003, about 11.6 per cent of the cultivated area which came to 2,96,154 ha was under organic agriculture. The number of organic farms came to 18,576, which was 9.2 per cent of the total number of agriculture farms in Austria.

The conversion of cultivated land to organic method of agriculture was slow till 1990. The federal government introduced subsidies for organic conversion of cultivation in 1991.

The government programme of agri-environment of 1995 and the conversion of grass lands into organic farms gave a great fillip to the organic movement. There was a decrease in the conversions in 2000, but it increased substantially from 2001 to 2003. The decline was due to the difficulties in marketing the organic produces in the country. This was an indication that the emphasis on farm production was not enough to promote organic agriculture.

The increasing phase from 2001 to 2003 saw about 700 farms having an area of 30,000 ha being converted. This came from lands cultivating cereals and vegetables. One of the reasons for the change is attributed to the realization of the benefits of organic farming by the young farmers with a better concern for ecology and understanding of economics than their parents. The organic farmers belonged to a number of associations and the latter had different approaches/objectives, regional differences and marketing techniques. About 64.5 per cent of the organic farmers were members of these associations, which functioned under two umbrella organizations, named ARGE and OelG. These associations have their own production and quality standards, which in many cases are more stringent, than those of the State. Presently there are 11 major organic farming associations in Austria and while 6 of them come under ARGE, the remaining five and 4 marketing associations are members of OelG, the latter being the one giving importance of marketing too in the promotion of organic agriculture in Austria.

The concentration of organic farms is in the regions with large areas of grass land. It is easier for the conversion of the grass land to organic farming as they used to be cultivated extensively. Grass land farms constitute about 66 per cent of the total organic farms. The average size of an organic farm in Austria is 16 ha. Certification of organic farms in Austria is done with the provincial government and a contract between the farm and an inspection agency is essential for registration. An approved organic label is available to guarantee the quality of the organic products. The conditions to be fulfilled by an inspection agency for accreditation are very tough in Austria. 17,500 registered organic farms receive government subsidies and about 50 per cent of it is born by the European Union and balance by the federal and provincial governments.

The market value of Austrian organic products was 2.18 billion Euro in 2000 which was 3 per cent of the total food market. About 70 per cent of these products is distributed by the conventional food chains. Now, the distribution of organic foods through the supermarket chains is gaining momentum. It is reported that the supermarkets selling organic products enjoy an improved image among their customers. Organic farms' cooperatives also market organic products. Cooperative ventures between the organic farmers and hospitals, nursing homes, etc. have come up to sell organic products. Home delivery services and organic shops are also established, particularly in large cities.

Agricultural Schools and Colleges offer training programmes/ courses in organic farming and promote scientific research. The Institute for Organic Farming at the University of Vienna is a major centre for research in organic agriculture.

Developmient of a strategy for co-operation among the organic farmers, their associations, processors and traders is considered a major challenge in the promotion of organic farming in Austria.

ORGANIC FARMING IN GERMANY

Germany, where the IFOAM is located at Tholey-Theley since 1987, is a leading nation in the adoption of organic farming in Europe. The IFOAM has its Regional Group founded in 1991 with the German speaking nations as members. Germany has about 3 per cent of its total agricultural area under organic farming against 2 per cent of the European countries as a whole. The country has also a large number of processors and importers of organic produces. So, it would be useful to examine the organic farming movement in that country in some detail.

The history and development of organic agriculture in Germany begins from the days of bio-dynamic farming initiated in 1924. Bio-dynamic theory of farming treats farm as a living being which responds to non-material influences, called dynamic forces. The proponents of bio-dynamic farming believe that these forces can be enhanced by bio-dynamic preparations. Even certification for bio-dynamic agriculture started in 1954 in the country. Organic-bio-dynamic farming was developed in Switzerland in the early twentieth century. Hans Mueller and his wife Maria Mueller were the spirits behind this movement in Switzerland. Mueller had developed the method during the 1950s and the theoretical basis was provided by Hans-Peter Rusch, the German microbiologist and medical doctor.

The damage inflicted to the environment by the modern farming system was clear by the end of 1960s. The ill effects of pollution were obvious and a producer organisation, named Bioland was established in 1971 in Germany. Another organisation for the organic cultivation of fruits and vegetables had been founded in 1961. A federation for organic farming associations was established in 1988 consisting of six producer organisations. A common basic standard for organic agriculture was laid down in 1984 as a framework for individual organization.

Organic farming spread very fast in Germany during the subsequent period. State funding for organic agriculture was available from 1989. The unification of both the Germanys in 1990 saw fast growth of organic farming in the former East German areas. The trade sector felt the absence of a German logo for organic products and the seal 'Biosiegel' was launched by the federal government. The German government has put the goal of bringing 20 per cent of the agricultural land under organic farming by 2010 from the 3 per cent of 2002. A series of measures were introduced to achieve the goal. These steps are under the EU programme of support to organic agriculture and the federal programme for organic farming.

There are eight organic producer organisations in Germany and they represent about 60 per cent of the organic farming community. All these associations have the common organic seals for labelling the certified products, both primary and processed. The organic producers, processors and traders of the countiy have formed a common organization aiming at the propagation of achievements of organic production, protection of environment etc.

There are several other organisations functioning in relation to organic farming in Germany. SOL, FiBL, Schweisl\irh Foundation, Foundation Future Agriculture and Society for Ecological Animal Husbandry are some of the institutions contributing to the spread of organic farming in Germany. Organic farming has gained acceptance in the former East Germany compared to the West. The reasons are the substantial changes in the German agriculture after the reunification, existence of large conservation areas in which organic farming can well be fitted into. Large areas in the eastern part are designated as disadvantaged and incentives are provided there to go for organic farming. The arguments for and against organic farming between the organic and conventional farmers are intense in the western parts of Germany and this has to some extent slowed down the spread of the organic agriculture there. However, government regulations and the support of European Union Regulations have made the organic agriculture accepted in the country.

The standards and certification process for organic farming are stricter in Germany than in other European Union countries. The German regulations restrict certification to the whole farm and the processing standards are tough. There are private and state standards and the inspections are done according to both of them. The use of genetically engineered organisms and techniques are banned in organic agriculture. There are more than 20 inspection bodies and the 22 states of the country control them. The estimated organic production in Germany.

The German organic food market was worth about 3 billion Euro in 2000. Both the federal and state governments extend financial support to organic farming. Farmers are encouraged to adopt organic methods from 1989 and those who adopt production methods compatible with the protection of environment are given assistance. In addition to the direct subsidies, marketing assistance is also provided to the organic farms. Subsidies for producers' marketing organisations, processing and market promotion are available.

Advisory services are provided to the farmers for organic farming. Advisory services were organised by the farmers themselves during the initial period of the introduction of organic farming in Germany. Now producer associations with governmental financial support offer such services. Farmers themselves hire advisors with state support and official advisors are also available.

Research and training facilities are existing in almost all the German agricultural universities and technical colleges in organic farming. Degree courses are offered in organic agriculture and organic farming research is also funded by the state. A state research institute for organic farming was established in 2000.

ORGANIC FARMING IN INDIA

The need for organic farming in India arises from the unsustaina-bility of agriculture production and the damage caused to ecology through the conventional farming practices.

The present system of agriculture which we call 'conventional' and practiced the world over evolved in the western nations as a product of their socio-economic environment which promoted an over riding quest for accumulation of wealth. This method of farming adopted by other countries is inherently self destructive and unsustainable.

The modern farming is highly perfected by the Americans who dispossessed the natives of their farms right from the early period of the new settlers in US. The large farms appropriated by the immigrants required machines to do the large scale cultural operations. These machines needed large amount of fossil fuels besides forcing the farmers to raise the same crops again and again, in order to utilize these machines to their optimum capacities. The result was the reduction of bio-diversity and labour. The high cost of the machines necessitated high profits, which in turn put pressure to raise productivity. Then, only those crops with high productivity were cultivated which needed increased quantities of fertilizers and pesticides. Increasing use of pesticides resulted in the damage to environment and increased resistance of insects to them. Pesticides harmed useful organisms in the soil.

The monoculture of high yielding seeds required external inputs of chemical fertilizers. The fertilizers also destroy soil organisms. They damage the rhizobia that fix nitrogen and other micro organisms that make phosphates available to plants. The long term effect was reduction of crop yields. The damaged soil was easily eroded by wind and water. The eroding soil needed use of continuously increasing quantities of fertilizers, much of which was washed/ leached into surface and underground water sources.

The theme of consumer welfare has become central in the economic activities in the developed countries in the world. Sustainable agriculture based on technologies that combine increased production with improved environmental protection has been accepted as absolutely essential for the maximization of the consumer welfare. The consumers are increasingly concerned about the quality of the products they consume and food safety has become a crucial requirement. Safety, quality and hygienic standards are increasingly being made strict.

The mad cow disease and the question of genetically modified food production are the recent instances, which made the countries to tighten the laws. Mycotoxln contamination, unacceptable levels of pesticide residues and environment degradation are the problems on which the attention is centred. Keeping the interests of the consumers, the European Union has taken tough measures including criminal prosecution to ensure food safety. Another area to increase the consumer welfare is promotion of the eco-friendly methods in agriculture. No-till, or conservation agriculture, lower input approaches of integrated pest or nutrient management and organic farming are some of them.

The Indian agriculture switched over to the conventional system of production on the advent of the green revolution in the 1970s. The change was in the national interest which suffered set backs because of the country's over dependence on the foreign food sources. The national determination was so intense that all the attention was focused on the increase in agriculture production.

The agriculture and allied sectors in India provide employment to 65 per cent of the workers and accounts for 30 per cent of the national income. The growth of population and the increase in income will lead to a rise in demand for foodgrains as also for the agricultural raw materials for industry in the future. The area under cultivation, obviously, cannot be increased and the present 140 million hectares will have to meet the future increases in such demands. There is a strong reason for even a decline in the cultivated area because of the urbanization and industrialization, which in turn will exert much pressure on the existing, cropped area. Science and technology have helped man to increase agricultural production from the natural resources like land. But the realization that this has been achieved at the cost of the nature and environment, which support the human life itself, is becoming clear. It has been fully evident that the present pattern of economic development, which ignores the ecology and environment, cannot sustain the achievement of man without substantial erosion of the factors that support the life system of all living things on the Earth. The evidence of the ill effects of development is well documented. As said earlier, we in India have to be concerned much more than any other nation of the world as agriculture is the source of livelihood of more than 6-7 million of our people and it is the foundation of the economic development of the country.

There were times when people lived close to nature with access to flora and fauna in healthier and cleaner surroundings. One has to look back at our present metropolitan cities or other large towns before the past fifty years as recorded in history/memmories of the present elder generation to see the striking differences in the surroundings in which the people lived there. Land, water and air, the most fundamental resources supporting the human life, have degraded into such an extent that they now constitute a threat to the livelihood of millions of people in the countiy.

Ecological and environmental effects have been highly publicised all over the world. Many times, these analysis have taken the shape of doomsday forecasts. Powerful interests in the developed western countries have also politicised these issues to take advantage of the poor nations of the world. Efforts to impose trade restrictions on the plea of environment protection are a direct result of these campaigns. But we have to recognize that the abysmal level to which we have degraded our resources,requires immediate remedial measures without terming the demand for them as the ploys of the rich nations to exploit the poor. Another turn of the events has been the blame game for ecological problems stated at the Earth Summit and other international conferences.

The developed countries, it is true, are to a great extent instrumental to degrade the environment. However, the poorer countries of the world including India cannot delay or ignore the need for remedial measures, which are to be effectively implemented. We cannot gloss over the fact that we have also contributed to the degradation of ecology; look at the droughts and floods, disappearance of forests, high noise level and air pollution in the cities which are our own creations.

Organically cultivated soils are relatively better attuned to withstand water stress and nutrient loss. Their potential to counter soil degradation is high and several experiments in arid areas reveal that organic farming may help to combat desertification. It is reported that about 70 hectares of desert in Egypt could be converted into fertile soil supporting livestock through organic and biodynamic practices. India, which has some areas of semi-arid and arid nature, can benefit from the experiment.

The organic agriculture movement in India received inspiration and assistance from IFOAM which has about 600 organizational members from 120 countries. All India Federation of Organic Farming is a member of IFOAM and consists of a number of NGOs, farmers' organisations, promotional bodies and institutions.

The national productivity of many of the cereal crops, millets, oilseeds, pulses and horticultural crops continues to be one of the lowest in the world in spite of the green revolution. The fertilizer and pesticide consumption has increased manifold; but this trend has not been reflected in the crop productivity to that extent. The country's farming sector has started showing indications of reversing the rising productivity as against the increasing trend of input use.

The unsustainability of Indian agriculture is caused by the modern farming methods which have badly affected/damaged production resources and the environment.

Affects of Modern Farming Technology

The role of agriculture in economic development in an agrarian country like India is a pre-dominant one. Agriculture provides food for more than 1 billion people and yields raw materials for agro-based industries. Agricultural exports earn foreign exchange. Modernization of Indian agriculture began during the mid-sixties which resulted in the green revolution making the country a foodgrain surplus nation from a deficit one depending on food imports. Modern agriculture is based on the use of high yielding varieties of seeds, chemical fertilizers, irrigation water, pesticides, etc., and also on the adoption of multiple cropping systems with the extension of area under cultivation. But it also put severe pressure on natural resources like, land and water.

However, given the continuous growth of modern technology along with the intensive use of natural resources, many of them of non renewable, it is felt that agriculture cannot be sustainable in future because of the adverse changes being caused to the environment and the ecosystem. The environmental non-degradable nature of the agricultural development and its ecological balance have been studied in relation to the modem Indian farming system by experts which shows exploitation of land and water for agriculture, and the excessive use of chemicals.

Chemical Contamination and Fertilizers

Consumption of chemical fertilizers {N,P,K) has been increasing in India during the past thirty years at a rate of almost half a million tonnes on an average, a year. It was only 13.13 kg/ha in 1970-71, 31.83 kg/ha in 1980-81 and 74.81 kg/ha in 1995-96. It shot up to about 96 kg/ha during 1999-2000. The consumption of fertilizers in India from 1970-71 to 2001-02. The present use of about 96 kg of fertilizers per ha in India appears to be modest compared to the advanced countries. Currently about 80 per cent of the fertilizer is consumed in only about 120 districts constituting less than 33 per cent of the gross cultivating area. Experts point out that the efficiency of fertilizer use in India is only 30-35 per cent as the balance 65-70 per cent reaches the under ground water. The intensity of their use in a few regions and a few crops are causes of serious concern to human health, soil, water, environment and thus to the sustainability of agriculture production in the country.

It is true that the increasing use of fertilizer at high rates has boosted agricultural production in the country. But it has also caused adverse impact on soil and water as well as environment. Several studies on the effects of high level of fertilizer application on soil health have confirmed the adverse impacts.

Both drinking and irrigation water wells in large numbers have been found contaminated with nitrates, some of them are having even 45 mg per litre, well above the safe level.

Long term continuous use of high doses of chemical fertilizers badly affects the physical, chemical and biological properties of the soil. A study at the University of Agricultural Sciences, Bangalore confirmed the deterioration of soil health because of the reduction in water holding capacity, soil pH, organic carbon content and the availability of trace elements such as zinc in case of ragi crop even with the application of normal doses of fertilizer in the long run. In the long run, increasing nitrogenous fertilizer use leads to the accumulation of nitrates in the soil. The application of sulphatic fertilizers leaves sulphates in the soil. Rainfall and excessive use of irrigation water cause these chemicals to change the alkaline or acidic nature of the soil. The nitrates go to the rivers, wells, lakes etc. And also leak into the drainage system which goes into the drinking water contaminating the environment. It also causes depletion of the ozone layer adding to the global warming. Use of nitrogen in the form of ammonium sulphate in the rice crop emanates ammonia polluting the atmosphere. The heavy metals present in the fertilizers and sewage sludge leach into ground water.

Chemical Pesticides

The use of chemical pesticides began with the discovery of toxicological properties of DDT and HCH during the Second World War. Many chlorinated hydrocarbon insecticides like aldrin, dieldrin, toxaphane, chlordane, endosulfan, etc. came into the market during the second half of the last century. Simultaneously, organophosphate and carbonate compounds were employed in agriculture. A new group of insecticides, such as premethrin, cypermethrin, fenalerate, etc. which were effective at low doses came into being in the 1970s.

The use of pesticides has helped in increasing agriculture production and also led to the development of resistance in pests, contamination of the environment and resurgence of many pests. There are about 1000 agrochemicals in use in the world over. India accounts for about 3.7 per cent of the total world consumption. At present, our consumption is about 90,000 tonnes of plant protection chemicals. It comes to about 500 grams per ha compared to 10-12 kg/ha in Japan and 5 kg/ha in Europe. However, the use of pesticides in India is uneven like the fertilizers. While in cotton it is about 3 to 4 kg/ha, in pulses it comes to below 500 grams/ha.

Agricultural chemicals have become a major input in Indian agriculture with the increasing demand for food, feed and fibre. The pesticide consumption was about 2000 tonnes annually during the 1950s. India happens to be the second largest manufacturer of pesticides in Asia after Japan. It is also of interest to know that in spite of increased consumption of plant protection chemicals, the produce loss due to insects and pests increased by 5 times during the period from 1988 to 1995.

Increasing application of fertilizer also leads to increasing use of pesticides to control pests and diseases. The trend of increasing fertilizer use also compels the farmers to enhance the use of pesticides as well. the use of fertilizers in increasing amounts leads to growth of weeds and in the process of weedicide use many plants growing nearby also get killed, which reduces the biodiversity. Meanwhile, the weeds also develop resistance to herbicides and the quest to formulate even powerful herbicides begins.

Consumption of pesticides increased from 24.32 thousand tonnes in 1970-71 to 75 thousand tonnes in 1990-91 and it slowed down during the subsequent period. Insects, pests and diseases like viral, bacterial and fungal affect the high yielding varieties of crops. Almost all pesticides are toxic in nature and pollute the environment leading to grave damage to ecology and human life itself. This indiscriminate use leaves toxic residues in foodgrans, fodder, vegetables, meat, milk, milk products, etc. besides in soil and water.

High doses of pesticides severely affect the aquatic animals, fish and the wild life. Insects develop resistance to insecticides in crops like cotton and in turn force the farmers to the excessive use of them. Cases of pesticide poisoning and human and animal deaths are also reported. Pesticides irritate the skin and the respiratory system in the humans gets damaged.

It was found that all water bodies like, rivers, canals, lakes, tanks and ponds and also the costal water were contaminated with high amounts of DDT, HCH and other organochlorine pesticides. River water is seen as more contaminated than other water sources.

Contamination of drinking water with DDT and HCH is reported from different states. Since the concentrations of contaminants are higher than MRL values fixed by the Environmental Protection Agency, the seriousness of the problem can be gauged.

Pesticides also contaminate animal feeds and fodder. Green fodder, paddy and wheat straw contain residues of DDT and HCH. Several studies have confirmed this trend. Milk and milk products are also affected by the pesticide use. Both bovine and human milk showed high levels of pesticide contamination. The sources of contamination of bovine milk are traced to the fodder and feed concentrates and in case of the human milk, the consumption of contaminated food by the lactating mothers is reported to be the reason.

Infant formula/baby milk powders also showed DDT and HCH contamination level ranging from 94 to 100 per cent. Butter and ghee, the other animal products revealed high contamination levels in many parts of the country.

Cereals like wheat and rice were seen contaminated highly by pesticides like, DDT, HCH and malathion. The case of vegetable, vegetable oil, honey, fish etc. is also not different as they too have unacceptable high pesticide residue content levels.

The adverse effect of pesticide contamination on humans in India is understood from the study of dietary intake. Such studies, although a few in number, have confirmed high levels of pesticides contamination which come to more than 3 to 5 times than the agriculturally developed countries.

The daily intake of pesticide per individual is estimated to be about 0.51 milli-grams which is above the accepted level. The Indian Institute of Horticulture Research has reported contamination of 50 per cent of the fruits and vegetables sold in the Bangalore market with the residues of DDT and HCH. Use of herbicides over a period results in the shift of the weed flora. The weeds of minor importance, often, become major weeds. Repeated application of weedicides helps the development of resistance in weed at alarming proportions. The remedy recommended is rotation of herbicides or the use of other herbicides. Any way, the end result is contamination of ground water and soils inflicting damage on environment.

The number of herbicides registered in India comes to about 28 in 1997-98 which was only 10 before 10 years. This is often compared to about 300 herbicides available in the North America. There are only 10 herbicides manufactured in the country and the herbicides consumption was about 6000 Tonnes during 1994-95. It is reported that the level of herbicide use in rice, wheat, and tea in India is almost the same that of the world at large. Sugarcane, soyabean, groundnut, coffee, cotton, onion and potato are the other crops, which find widespread application of herbicides in India. The contamination of water, air and soil with toxic synthetic fertilizers, pesticides and herbicides leads to increasing deaths of many creatures, and to human illness and mortality.

The end result is loss of biodiversity and natural harmony, increased expenditure to purify water, air, etc. The toxins in the food crops cannot be removed and the threat to human existence itself seems to be real.

The firms engaged in the manufacture and supply of agricultural inputs have a vested interest in keeping the input use increasing. Besides, they influence the government policy towards agriculture. '

Vigourous Growth and High Yields of Crops

Water is one of the important inputs for the vigourous growth and high yields of crops. The modernisation of Indian agriculture has resulted in the increased use of irrigation water. The area under irrigation has grown substantially during the past three decades. The gross irrigated area in the country. The gross irrigated area of 38.18 million ha in 1970-71 increased to 49.73 million ha in 1980-81 and the next decade ending 1990-91 saw this further rising substantially to about 62.47 million ha. It increased to 72.78 million ha during 1997-98.

Heavy irrigation is necessary to get high production, as the new varieties cannot withstand water scarcity. This leads to salinity and water logging leaving the land uncultivable. Over exploitation of underground water is another effect. When water table falls, increasing energy will be required to lift water for irrigation.

Irrigation is necessary for the vigourous growth and high yields of crops in the modern method of cultivation. Many of the crops, particularly the rice and wheat high yielding varieties need more irrigation water than the traditional varieties.

The area under irrigation in the country is only about 35 per cent and the remaining is still dependent on rains. So, there is a necessity to use irrigation water judiciously. Its excessive use results in severe ecological dangers like water logging of vast cultivated areas by seepages from canals. The loss of water through seepages and evaporation is estimated to be about 38 per cent. Flooding also results in run off and leaching losses of fertilizer nutrients, pesticides and soil particles. Excessive use of canal water makes the field vulnerable to soil erosion.

The excessive irrigation in certain areas results in wastage as evident from the water logging of vast cultivated areas caused by the seepage from many virater sources.

Water logging is harmful to the soil. Seepage of canal water leads to salts present in the lowest layer of soil come up to the surface and the soil may turn alkaline or saline. Dams and multipurpose projects degrade the soil in the command area due to soil salinity and water logging. The chambel region in Rajasthan and Madhya Pradesh, the command area of the Bhakra Nangal, etc are the examples of water logging created by huge water irrigation projects.

Crops irrigated by sewage water have adverse effects on the health of the human population consuming the produce. The workers work on these farms also face health hazards.

Depletion of Energy Resources in Chemical Fertilizers

Chemical fertilizers, pesticides, herbicides, etc are manufactured using the non-renewable materials like the fossil fuels. The global demand for oil and natural gas is increasing and thus the price of the inputs to agriculture is bound to rise. India's petroleum resources, which presently meet only about 30-35 per cent of the consumption demand, are under pressure. Increasing demand for chemicals and energy in agriculture sector will have affects on our energy sources.

The investments in agriculture have to be increased to meet the rising input costs and larger areas are brought under farming to earn profits. Large farms have to transport the produce to distant areas. Again, energy will be required for transportation, processing and packaging.

The rice-wheat cropping pattern and the cultivation of crops like sugarcane require high irrigation, which results in the depletion of water level. Singh and Singh found that the water level in the states of Punjab and Haryana had gone down by 0.3 to 1 m per annum during a period of 10 years due to the excessive use of water for paddy crop.

Growth Process of Input-Output Lmbalance

A crop, in its growth process, incorporates a part of the soil fertility into the parts of the plant. The roots remain in the soil. The leaves and stems are fed to the cattle/burnt as fuel/directly returned to the soil. The consumed part by cattle and human also go back to the soil. The practice of commercial farming leads to continuous export of the soil fertility to outside the farming areas as the organic matter leaves the locality. The soil nutrients in the form of farm produces continue to be exported. The import of chemical fertilizers cannot compensate the loss of soil nutrients through exports. The soil becomes powdery and gets eroded by wind or rain. If the harvests are exported from the country, the loss is higher.

The Earth can produce only a limited amount of biomass from a given area. If man tries to extracj: more, the system degenerates. If we realize that we live in a closed eco-system and act in tune, the resources can be used iri a sustainable way. The maintenance of a dynamic equilibrium balances the inputs and outputs in a region. If agriculture becomes unsustainable, no society can survive for a long time.

Not only the intensive cultivation through the use of technological inputs, but also the extensive crop production through Increase in the area under cultivation has been an important aspect of modern agriculture seen in India. Increasingly areas under forests are brought under plough along with the marginal, sub-marginal and undulating land. The net sown area was 140 million ha in 1970-71 and stood at 142 million ha at the end of 1997-98.

The genetic base of crops is very important and a reduction of genetic diversity leads to the emergence of pests on a large scale. Farmers, in olden times, apart from' using the crop rotation methods to maintain the soil fertility also relied on the genetic means to increase crop production.

Relying exclusively on nation's own reserves of fertility and immunology, the farming community by evolving trial and error methods discovered. hybrid varieties of crops by crossing the related strains. These crosses were from the same environment and no violence was used to separate them from nature by maintaining the ecological balance.

The high yielding varieties of crops are the crosses from different environments and distantly related strains. The high yielding rice variety got by the crossing between the dwarf and non-dwarf varieties has major genetic weaknesses. The dwarf gene is susceptible to pest and viral attacks and the seed cannot manifest its potential without chemical fertilizer. Thus, an artificial environment has to be created for the growth of the crop.

Thus synthetic fertilizers supplant natural fertility, which results in larger population of pests.

The new technology adopted then depends upon the replacement of the local/traditional varieties of seeds. But this results in the reduction of genetic diversity and increase in genetic erosion. These modern technologies are but the result of clever manipulations of nature's genes.

The productivity of cereals, millets, oilseeds, pulses and plantation crops is very low in comparison with those in other countries in the world. This is in spite of our success in improving the quality of seeds and adoption of efficient technology.

The impact of green revolution is showing signs of weakness and production appears to have decreased even after an increase in the inputs used. The production of foodgrains in the country increased very substantially during 1960 to 1980 to reach 160 million Tonnes from 60 million tonnes. But the decade ending 1990 and 2000 did not witness such increases and the attainment of the targeted production of 240 million tonnes to meet the demand of the population by 2010 seems to be difficult.

The reasons attributed to the low productivity are the drastic reduction in soil nutrients in the areas where fertilizer is used intensively in which the organic matter is not supplemented.

ORGANIC FARMING IN DEVELOPMENT COUNTRIES

Sustainable development has caught the imagination and action all over the world for more than a decade. Sustainable agriculture is necessary to attain the goal of sustainable development. Food and Agriculture Organization, sustainable agriculture "is the successful management of resources for agriculture to satisfy changing human needs while maintaining or enhancing the quality of environment and conserving natural resources". All definitions of sustainable agriculture lay great emphasis on maintaining an agriculture growth rate, which can meet the demand for food of all living things without draining the basic resources.

Organic farming is one of the several approaches found to meet the objectives of sustainable agriculture. Many techniques used in organic farming like inter-cropping, mulching and integration of crops and livestock are not alien to various agriculture systems including the traditional agriculture practiced in old countries like India.

However, organic farming is based on various laws and certification programmes, which prohibit the use of almost all synthetic inputs, and health of the soil is recognised as the central theme of the method.

Adverse effects of modern agricultural practices not only on the farm but also on the health of all living things and thus on the environment have been well documented all over the world. Application of technology, particularly in terms of the use of chemical fertilizers and pesticides all around us has persuaded people to think aloud. Their negative effects on the environment are manifested through soil erosion, water shortages, salination, soil contamination, genetic erosion, etc.

Organic farming is one of the widely used methods, which is thought of as the best alternative to avoid the ill effects of chemical farming. There are several definitions of organic farming and the one given by the US Department of Agriculture is considered the most coherent and stringent. It is defined as 'a system that is designed and maintained to produce agricultural products by the use of methods and substances that maintain the integrity of organic agricultural products until they reach the consumer.

This is accomplished by using substances, to fulfill any specific fluctuation within the system so as to maintain long term soil biological activity, ensure effective peak management, recycle wastes to return nutrients to the land, provide attentive care for farrn animals and handle the agricultural products without the use of extraneous synthetic additives or processing in accordance with the act and the regulations in this part'. The origin of organic farming goes back, in its recent history, to 1940s. During this period, the path breaking literature on the subject published by J.I. Rodale in the United States, Lady Balfour in England and Sir Albert Howard in India contributed to the cause of organic farming.

The farming being practiced for the last three decades in India has increasingly been found non-sustainable. The system is oriented towards high production without much concern for ecology and the very existence of man himself.

RELEVANCE OF ORGANIC FARMING FOR INDIA

There are three categories of opinions about the relevance of organic farming for India. The first one simply dismisses it as a fad or craze. The second category, which includes many farmers and scientists, opines that there are merits in the organic farming but we should proceed cautiously considering the national needs and conditions in which Indian agriculture functions. They are fully aware of the environmental problems created by the conventional farming. But many of them believe that yields are lower in organic cultivation during the initial period and also the cost of labour tends to increase therein. The third one is all for organic farming and advocates its adoption wholeheartedly. They think that tomorrow's ecology is more important than today's conventional farm benefits.

However, among many a major reservation, the profitability of organic farming vis a vis conventional farming, is the crucial one from the point of view of the Indian farmers, particularly the small and marginal. Organic farming involves management of the agro-eco system as autonomous, based on the capacity of the soil in the given local climatic conditions. In spite of the ridicule poured out on organic farming by many, it has come to stay and is spreading steadily but slowly all over the world. India has been very slow to adopt it but it has made Inroads into our conventional farming system. One advantage we have here is the fact that the farming techniques practiced in this country before the advent of the green revolution were basically eco-lriendly and they have not faded away from the memories of the present elder generation of our farming community.

India's options in finding out an alternative method to the conventional farming are limited. The World Trade Organization deadline for banning exports of vegetables and horticulture products based on inorganic farming will expire in 2005. Sensing the importance, the Central and state governments have taken several initiatives to popularise organic farming in the country.

RELEVANCE OF ORGANIC FARMING

The relevance and need for an eco-friendly alternative farming system arose from the ill effects of the chemical farming practices adopted w ^ orldwide during the second half of the last century. The methods of farming evolved and adopted by our forefathers for centuries were less injurious to the environment. People began to think of various alternative farming systems based on the protection of environment which in turn would increase the welfare of the humankind by various ways like clean and healthy foods, an ecology which is condusive to the survival of all the living and non-living things, low use of the non-renewable energy sources, etc. Many systems of farming came out of the efforts of many experts and laymen.

However, organic farming is considered to be the best among all of them because of its scientific approach and wider acceptance all over the world. The negative effects of modem chemical based farming system were first experienced by those countries, which introduced it initially. So, naturally, it was in those countries organic farming was adopted in relatively large scales. There are very large organisations promoting the organic farming movcmcnt in European countries, America and Australia etc. These organisations the International Federation of Organic Agriculture Movements and Greenpeace have studied the problems of the chemical farming methods and compared the benefits accruing to the organic faming with the former. Organic farming movements have since spread to Asia and Africa too.

IFOAM was founded in France in 1972. It spearheads and coordinates organic farming efforts the world over by promoting organic agriculture as an environment friendly and sustaining method. It focuses on organic farming by highlighting the minimum pollution and low use of non-renewable natural resources through this method. It has about 600 organisational members spread over about 120 countries including India.

IFOAM undertakes a wide range of activities related to organic farming such as exchanging knowledge and thoughts among its members; representation of the movement in governmental, administrative and policy making forums in the national and international arena; updating of production, processing and trading standards; formulation and coordination of research projects; and holding of international conferences and seminars.

IFOAM participates in the activities related to organic farming under the auspices of the United Nations and keeps active contacts with several international NGOs.

The Food and Agriculture Organisation of the United Nations provides support to organic farming in the member countries. It also attempts the harmonization of national organic standards, which is absolutely essential to increase international trade in organic products. The FAO has, in association with the World Health Organisation, evolved the Codex Alimentarius for organic products.

Organic farming has several advantages over the conventional one apart from the protection of both the environment and human health. Improved soil fertility, better water quality, prevention of soil erosion, generation of rural employment, etc. are some of them.

The concept of quality food has undergone a drastic change over the past few decades. It does give emphasis on the characteristics of the end product, but the process and method of production and transport are now considered equally important.

Not only the importers but also the domestic retailers have their own quality specifications or standards, which in many cases are tougher than those of the government regulations. Consumers have become health conscious and are willing to pay for the clean, healthy and natural food. Many developed countries have various support programmes to help organic farming with financial incentives and technical guidance.

The organic food market in the world has grown rapidly in the past decade. International trade in organic foods showed an annual growth rate of about 20-

22 per cent during this period. Many retail chains and supermarkets in advanced countries are accorded with 'green status' to sell organic foods. The organic food processing industry is considered nature friendly and thus encouraged. The size of the organic food markets in some developed countries in 2000. The US food market tops with a value of $8.0 billion, followed by Germany and Japan with $2.5 billion each while France accounts for $1.25 billion. Moreover, the organic foods command higher prices which often range from 10 to 100 per cent more than the conventional ones.

3

Tropical Crops and Tissue Culture

HORTICULTURE RESEARCH

India has a wide variety of climate and soil on which a large range of horticultural crops such as, fruits; vegetables, potato and other tropical tuber crops; ornamental, medicinal and aromatic plants; plantation crops; spices, cashew and cocoa are grown.

After attaining independence in 1947, major emphasis was laid on achieving self sufficiency in food production. Development of high yielding wheat varieties and high production technologies and their adoption in areas of assured irrigation paved the way towards food security ushering in green revolution in the sixties. It, however, gradually became clear that horticultural crops for which the Indian topography and agro climates are well suited is an ideal method of achieving sustainability of small holdings, increasing employment, improving environment, providing an enormous export potential and above all achieving nutritional security.

As a result, due emphasis on diversification to horticultural crops was given only during the last one decade.

RESEARCH INFRASTRUCTURE

The Indian Council of Agricultural Research is the premier agency which pioneered systematic research on agricultural crops in the country. Horticulture research in India received very little attention till the 3rd Five Year Plan.

The establishment of the Indian Institute of Horticultural Research at Bangalore and starting of eight All India Coordinated Crop Improvement Projects to cover different horticultural crops was a landmark in the history of horticulture in 4th Five Year Plan (1969-74). Rapid expansion of infrastructure took place in 7th and 8th Plans.

Today, the horticultural research in the country is being carried out at eight ICAR institutes (with 26 regional stations), 10 National Research Centres (on major crops) and a Project Directorate on Vegetable crops. Area specific, multi-disciplinary research is also being conducted under 14 All India

Coordinated Research Projects each on Tropical, Subtropical and Arid Fruits; Vegetables, Potato, Tuber Crops and Mushrooms; Ornamental Crops, Medicinal and Aromatic crops; Palms, Cashew, Spices and Betel vine; and Post Harvest Technology at 215 centres located at various research Institutes, and State Agricultural Universities.

In addition, four net work projects each on hybrid research in vegetable crops, drip, irrigation in perennial horticultural crops, protected cultivation of ornamental crops and *Phytophthora* diseases of horticulture crops are now in operation. Research on horticulture is also being undertaken at several multi-crop, multi-disciplinary Institutes.

Departments of Horticulture in 24 Agricultural Universities, one deemed to be University and one full fledged University of Horticulture and Forestry are also engaged in horticultural research. Besides 280 adhoc schemes supported from Agriculture Produce Cess Fund and a number of foreign-aided projects have also been in operation on specific problems of different horticulture crops. As a result, the country now has a sound research infrastructure in horticulture to meet the growing needs and expectations of the fast developing horticulture industry.

Budgetary Support

The investment in horticulture research by the ICAR in the Central sector has increased significantly in the last two Plans. The Plan allocation for horticultural crops started in 4th Plan (1969-74) with a modest allocation of ₹ 34.78 million and was enhanced to ₹ 319.56 million in the 7th Plan (1985-90) and to ₹ 1047 million in the 8th Plan (1992-97).

Non-Plan expenditure also increased from ₹ 73.55 million in the 5th Plan to ₹ 768 million in 8th Plan. Overall increase in Plan investment in 25 years has been of the order of 2775.21 per cent. The per cent budget allocation for horticulture research out of the total budget for agriculture research rose from 6.1 in 5th to 6.5, 6.67 and 7.7 in 6th, 7th and 8th five year plans, respectively. Similarly, expenditure for Central Sector Schemes of the Department of Agriculture and Cooperation for horticulture crop development also rose tremendously from ₹ 20.5 million (4th Plan) to ₹ 76.18 million (5th Plan), ₹ 146.37 million (6th Plan), ₹ 250 million (7th Plan) and ₹ 10,000 million (8th Five Years Plan).

Table. Budgetary Support (Million Rupees)

Five year plans	*Total for Agriculture Research*	*Share of Horticulture (%)*	*Total Development Support*
IV 1969-74	6105	N.A	20.50
V 1974-78	7292	6.10	76.18
VI 1980-85	10684	6.50	146.37
VII 1985-90	8445	6.67	250.00
VIII 1992-97	15165	7.70	10,000.00

Manpower

Nearly one sixth of the total strength of 5906 scientists working in ICAR is allocated for horticulture research in ICAR Institutes. Besides, approximately, 560 scientists are working in State Agricultural Universities in ICAR funded All India Coordinated Projects. In addition, a large number of scientists are working on horticultural crops in State Agricultural Universities.

Total Scientists in ICAR	5906
Scientists working in the Division of Horticulture	832
Scientists working in other ICAR Institutes	150
% of Total	15
Scientists working in SAUs in ICAR funded projects	560

Research Achievements

Several new crops have been introduced for commercial cultivation, *e.g.*:

- Kiwi fruit in sub-mountain areas of North India
- Olive in mid hills of North Western Himalayas
- Low chilling stone fruits in the North Western plains
- Oilpalm in coastal states of Karnataka, Andhra Pradesh, *etc*.
- Gherkin in south and west India
- Baby corn and sweet corn in certain specific pockets, and
- Broccoli, Brussels' sprouts, asparagus, celery, parsley near the cities.

Crop Improvement

- A large number of high yielding varieties developed in several horticultural crops *e.g.* fruits (76), vegetables (160), potato (29), other tuber crops (24), ornamental crops (300), palms (20), spices (51), cashew (33) and betel vine (1).
- First seedless variety of mango developed.
- 40 F_1 hybrids developed in brinjal, tomato, chillies, cauliflower, carrot, capsicum and muskmelon.
- Self incompatible lines in cauliflower, gynodioecious lines in cucumber and muskmelon, genetic male sterile lines in tomato and temperature tolerant strains of button mushroom developed.

Propagation of Quality Planting Material

- Standardized propagation technique for many fruits hitherto propagated by seed. *e.g., aonla, bael, ber,* black pepper, cardamom, cashew, cassia, cinnamon, clove, custard apple, jack fruit, *jamun,* nutmeg, sapota and walnut.
- Standardized Seed Plot Technique resulting in successful disease free potato seed production in the tropics and sub tropics of the country. Standardized method of micro-propagation and *in vitro* micro-tuber production in potato.

- Identification of suitable parental lines for production of True Potato Seed and Standardized technology for raising commercial crops.
- Micropropagation protocols developed in banana, black pepper, betel vine, cardamom, ginger and turmeric.
- Production of coconut hybrids through establishment of Seed gardens of Tall (T) x Dwarf (D) and D x T hybrids.
- Standardized rootstocks in citrus, grape and apple.

Agrotechniques

- Standardized high density plantations in banana, citrus, mango and pineapple and high production technology in several crops *e.g.*, pineapple, black pepper and cardamom.
- Year round production technology in tomato and "off season" cultivation of onion and cauliflower developed.
- Standardized use of several plant growth regulators and chemicals now commercially employed in production and quality improvement of horticultural crops *e.g.*, paclobutrazol for induction of flowering in mango; gibberellic acid for improving berry size and quality in grape; Maleic hydrazide for preventing sprouting in onion and potato; Dormex for hastening bud burst in grapes; and Boron and Calcium for changing flower cycle in some cucurbits.

Crop Protection

- Developed improved disease detection techniques such as ELISA and ISEM for improving seed quality and tissue culture technique for rapid multiplication of potato.
- Developed IPM for fruit borer in brinjal, diamond back moth in cabbage, thrips in chillies, phytophthora foot rot in black pepper, "Katte", rhizome rot in cardamom, rhizome rot in ginger, late blight and bacterial wilt in potato.
- Developed biological control measures for mealy bug in grape, fruit borer in tomato and okra.

Post Harvest Management

- Preharvest treatments to control post harvest losses in citrus, mango and grape standardized.
- Chemical treatment for regulation of ripening in mango, sapota and banana standardized.
- Optimum storage temperatures worked out for several fruits, vegetables and tuber crops.
- A mango harvester, fruit peeler, hand and pedal operator cassava chipping machine, harvesting tools (5-14 times efficient); Implements for mechanization of potato cultivation *e.g.*, oscillating tray type potato

grader, fertilizer application cum line marker, potato culti-ridger, soil crust breakers, potato digger and automatic potato planter/diggers developed.
- Low cost environment friendly storage system for fruits, vegetables, potato and onion developed.

TISSUE CULTURE TECHNIQUES

SOMACLONAL VARIATIONS

The production of plantlets by callus regeneration, cell suspensions, protoplast cultures, could present some deviations with regard to the mother plant. This is a way to increase the genetic variability. Associated with a selective pressure (stress to toxins, pH, salinity, cold,...) it's used to obtain resistant lines. Indeed, after regeneration, plants can express new potentialities rarely obtained another way. Stable and profitable variants are sale another way.

Stable and profitable variants are selected and introduced in breeding programmes. In 1976, a Pelargonium cv Velvet Rose was created by this technique. Later, other applications were described, as in sugarcane, tomato, red pepper,... Following Sibi (1994), extrachromosomic systems must be involved to explain the genetic behaviour of some new expressed tissue culture characteristics. She uses the term "epigenic" to evoke a whole of hereditary elements which belongs to the cytoplasmic or nuclear compartments, but are not transmitted by mendelian rules.

Haplodiploidisation

Many species are able to produce haploids through different *in vitro* techniques. The oldest was anther culture of androgenesis. To date, the results vary considerably from one species to another. Solanae (datura, tobacco, red pepper, eggplant, petunia), cereals (wheat, barley, rice, triticale, maize) or crucifers (soybean, cabbage) are species easy to regenerate by anther culture. To the contrary, tomatoes, leguminous, or Compositae are recalcitrant. Ovule culture, or gynogenesis, was a successful technique for Gerbera, beet, courgette,... However, the most common technique is pollination with irradiated pollen, in order to induce *in vivo* parthenogenesis. The oospheres developed in embryos without fertilisation are saved by embryo rescue.

The haploid plantlets (n chromosomes) are tree. The haploid plantlets (n chromosomes) are treated with colchicine to produce fertile homozygous lines, called "double haploid lines". This haplodiploidisation technique could give immediately new elite genotypes, hybrid parents after *in vitro* propagation, as asparagus supermales (MM), or useful genetic material to establish gene mapping. A good synthesis book concerning all these possibilities was published by Bajaj (1990).

PROTOPLAST CULTURE

Protoplasts are the smallest units able to regenerate a whole plant. Therefore protoplasts cultures can serve to enlarge genetic variability by introducing somaclonal variation. However, the main interest of protoplasts, is their capacity to fuse and to produce hybrids or cybrids, new organisms most often unknown in nature. Indeed, protoplasts allow to transgress the barrier of the botanical species or genus. Naked protoplasts can accept without rejection external elements: nuclei, cytoplasmic corganelles, liposomes,.... containing genetic information. The first protoplast fusion application was a cytoplasmic transfer from one genotype to another to induce male sterility (CMS) from mitochondrial origin. These male sterile hybrids are interesting to produce F1 hybrids. Another cytoplasmic transfer concerns the introduction of chloroplastic DNA to induce herbicide (atrazine) resistance, from Solanum nigrun duce herbicide (atrazine) resistance, from Solanum nigrun to tomato for instance.

Many intergeneric and interspecific hybridations are possible but not always profitable because these hybrids are too often infertile. This problem does not exist if we can vegetatively propagate the hybrids, as in the case of Citrus rootstocks. In Florida, Grosser (1990) was able to analyse several thousand Citrus hybrids.

CONTRIBUTIONS OF MOLECULAR BIOLOGY IN HORTICULTURE

More and more pathologists are using molecular probes for early detection of diseases. However, the main contributions of modern biotechnology remain probably in the hands of the plant breeders. Indeed, molecular markers are of great use in detecting desirable new genes, or to identify important QTL. This marker-assisted selection is very useful for many vegetable plants. Other applications are DNA - fingerprinting and genetic engineering. This recent technology aims to insert and to express new specific genes into a selected plants.

MARKER-ASSISTED SELECTION

Molecular markers and genome mapping will reach a large extension in the next breeding programmes of tomatoes, peas, cabbages, melons, potatoes,... It will be possible to improve the speed and the incorporation efficiency of desirable new genes by utilizing of closely linked selectable molecular markers genes by utilizing of closely linked selectable molecular markers. The number of backcross will be reduced. Tanksley *et al.* (1981) show that only with 12 markers, one on each tomato chromosome, the composition of the recurrent parent genome is similar to that observed in the absence of selection after the third backcross two years later. Therefore, it's easy to understand the high improvement that we can get with the 700 tomato markers already known in

1990 for this crop. The utilization of molecular markers could also identify important quantitative trait loci (QTL). Recently Foolad and Chen (1998) have identified 13 RAPD markers at eight genomic regions, that were associated with QTLs affecting salt tolerance during germination in tomato.

DNA Fingerprinting

The DNA polymorphism observed by RFLP, AFLP, or RAPD, allows cultivar identification of fruit trees, apple, citrus,... or other vegetatively propagated plants. Although some mutants, as coloured fruits, escape these identification techniques, some people would evaluate the genetic uniformity of regenerated plants in this way.

Genetic Engineering

Genetic engineering consists of introducing a foreign gene into a plant genome to create a new function. Or at the contrary, it could be to reduce or suppress an existing gene function, as in the antisense strategy.

1. *Historic:* In this field, progress were very rapid. In 1974 the pathogenicity of Agrobacterium tumefaciens, due to a large plasmid called Ti, was demonstrated. In 1977, Chilton *et al.* demonstrate a stable incorporation of plasmid Ti-DNA into higher plant cells. In 1983, the team of Van Montagu and Schell in Gent succeeded in producing a transgenic plant, and eleven years later the first transgenic cultivars were available in the market.
2. *Transgenic Plants with New Agronomic Traits:* Two different techniques are commonly used to introduce genetic information (DNA) inside cells, protected by pectocellulosic walls. A particle bombardment process, called biolistic, is used with success for monocots gynmosperm, squash, peas,...Today, this technique allows the bombardment of meristems and avoids the difficulties of regeneration. But the mediation of Agrobacterium remains till now the most routinely system to transform dicotyledonous plants.

   The main agronomic traits introduced in horticultural plants and already commercialised are Bt toxin and herbicide resistance. Other studies concern virus resistance, male sterility...
3. *Antisense Strategy:* Calgene created in 1994 the first commercial transgenic plant, a long shelf life tomato, by the suppression of polygalacturonase activity due to an antisense gene. However, t activity due to an antisense gene. However, this Flavr Savr tomato variety was removed from the trade 3 years later, because of its disease susceptibility and its lack of productivity !

   Later, other tomato varieties with long storage qualities were obtained by the utilization of an antisense RNA inhibition of ACC synthase or ACC oxydase, two ethylene precursors.

The antisense technique was also used to reduce the lignification of woody plants, by blocking the enzymes involved in the precursor of lignin biosynthesis. Another interesting application was the induction of white flowers in petunia and in different other ornamental plants by the suppression of chalcone synthase activity.

4. *Transmission of the New Traits:* These traits induced by the transferred DNA are transmitted to the progeny as a dominant Mendelian character. Nevertheless in some cases, transgenic plants with a strong gene expression can generate progenies with only a faint no gene expression. This problem is not clearly elucidate.
5. *Marker, Reporter, Promotor and Expression Genes:* To select the transformed cells, some marker genes, very often resistant to antibiotics or to herbicides, are attached to the coding sequence, as a promoter or an expression gene. Those allow the new gene to express in the whole plant or in a specific plant tissue.
   Some reporter genes are used to follow the evolution one.
   Some reporter genes are used to follow the evolution of the transformed cells. The most frequent reporter genes are the gene "gus" or the gene lux (luciferase).
6. *Horticultural Transgenic Crops already released in USA:* The Flavr Savr tomato was the first genetically engineered whole food approved for commercial sale in 1994. Four other transgenic tomatoes with a delayed ripening were approved later (1995 and 1996).

In 1995, potatoes with Bt genes and a squash cultivar resistant to two viruses were released. Two years later, another squash cultivar resistant to 3 viruses and a papaya line also resistant to viruses were approved. Two other horticultural crops are waiting the authorities approval: red hearted cichory (Radicchio) with male sterility and resistance to herbicide, and a tomato with Bt toxin.

It there are so few commercial transgenic plants in horticulture, it's probably due to the relatively low acreage of these horticultural crops in comparison with other agricultural crops. In 1997, in the world 5 million ha were planted with glyphosate resistant soybeans released by Monsanto alone.

Nowadays, it's true also that many transgenic plants exist in research laboratories, awaiting authorization for field testing, as they are very interesting model plants for learning physiology: lettuce with less nitrate by increasing nitrate reductase gene active with less nitrate by increasing nitrate reductase gene activity, a yeast ribonuclease gene to reduce viroid infection on potato, modification of lignin synthesis in plant to produce trees adapted to paper industry, or timber, or biomases production, regulation genes to control tree architecture,....

Situation of Biotechnologies in the World: Development and Risks

Situation in the Young Countries or in Developing Countries: A general use of *in vitro* micropropagation in the developing countries is now a reality, as it was predicted by Albert Sasson (1993). African countries for instance are producing potatoe, banana, cassava...

Even the production of transgenic plants is no longer a prerogative of the Northern countries. Different experiments are now starting in the South, specially in the international laboratories from CGIAR. Although CGIAR's expenditures on biotechnology for 1997 raise only 24.2 million US$ on an annual budget of 345 million US$ this Consultative Group on International Agricultural Research plays a very important role as contributor to agricultural research for developing countries.

The large international research centres IITA (Ibadan, Nigeria), CIP (Lima, Peru), CIAT (Cali, Columbia),... belong to the CGIAR. Their roles and NGO's are primordial. The genetic transformation their roles and NGO's are primordial. The genetic transformation of cassava is an excellent example. Its tuberous roots provide food for over 500 million people, mostly small-scale farmers. Unfortunately, till recently this integral plant for food security in developing countries has been recalcitrant to transformation approaches. However, thanks to the participation of international institutes located in cassava growing countries, CIAT and IITA with the help of four development associations from Swiss, UK, The Netherlands and USA, were able to initiate procotols for cassava transformation and regeneration.

Other networks, like the Cassava Biotechnology Network (CBN) previously described, exist. REDBIO, a technical cooperation network on plant biotechnology, is supported by FAO to promote a best use of scarce manpower, equipment and other resources in Latin America. On the other hand, more and more private laboratories exist in the developing countries which have a high scientific and technical level (China, Singapore, Taiwan, India, Brasil, Mexico, Chile,...). In these countries, European, American or Japanese agrochemical companies are investing. Also these countries will have their chance to develop their own technologies, and to export to the Northern markets.

Meanwhile, genetic engineering techniques are being applied mainly to crops which are important for g applied mainly to crops which are important for the industrialized world, not crops on which the world's hungry depend. Therefore, it is unrealistic when Monsanto writes "the experts said that biotechnological innovation will increase the crop productivity without occupation of new lands, saving tropical forest of best quality and animal habitat". However, the World Bank was promising in a 1997 publication Bioengineering of crops that" transgenic crops could improve food yield by up to 25 per cent in the developing countries and could help to feed an estimated additional three billion people over the next 30 years".

But, not all people are confident in this prospective. To the contrary, some are afraid that the poorest countries, where more than 700 million people are chronically undernourished will miss the biotechnological revolution.

Nevertheless, there is some hope to transfer to the developing countries high technology, which don't undermine the environmental and social network. So, the Agricultural Biotechnology for Sustainable Productivity Project (ABSP project) is managed by Michigan University. Four American universities, two research centres from France and USA, and two American private companies are also involved in this project. The ABSP's objectives are the reduction of losses due to pathogens and the reduction of losses due to pathogens and pests by using transgenic plants (sweet potatoes, potatoes, cucurbits, tomatoes) and by cloning commercial value-plants (bananas, pineapple, coffee).

Situation in the Northern Countries

1. *Mass Propagation:* In 1988, Pierik (1991) predicted a European *in vitro* propagation of 212 million plantlets/year. He was also predicting an important development of this activity for the future. Following Pierik this quantity accounted for only 5 per cent of the plants able to be propagated by vegetative means. However a few years later Pierik's prediction was not confirmed. The world production was estimated at 600 million. In USA, Zimmerman (1997) was recording a production of 121 million plantlets for a total of 110 laboratories, half produced in 11 laboratories, and 6 labs were propagating each more than 6 million plants per year.
   In Europe, only the Dutch production of micropropagated plants is precisely known. In 1990 the total production was 95.5 million. In 1995, the Dutch in country production reached only 53,7 million, whereas importations increased 77,3 million. (37 million from Poland and 17,1 million from India) (Pierik, data unpublished). These laboratory relocations to low-cost manpower countries are commonly observed, and put the question of the quality management.
2. *Transgenics:* The GMOs invade progressively all the US lands: 1.6 per cent in 1996, 15 per cent in 1997, 50 per cent in 2000..... It will represent an estimated market of 3,9 million US$ in 2003 ! The largestt agrochemical companies are merging to become giants of agrobusiness such as Novartis, Monsanto, Zeneca,...
   Following Monsanto in 1996, about 950 T insecticides were saved by introduction of resistant cotton, id est a net savings of 81 US$/ha for the grower. Glyphosate-resistant soybeans (4 million ha in 1997) allow a savomgs of 44 to 49 US$/ha. (Information from cultivar, suppl. n 436, Feb. 16, pp. 24-37, 1998).
3. *Perspectives, Limitations and Environmental Risks*: Ecological impact related with the introduction of GMOs is always an open debate.

> The application to agriculture of these new technologies certainly opens interesting perspectives, but also raises potential problems. The risk of crop transgene spreading has been demonstrated. A researcher of Clemson University in South Carolina reported "that in a population of wild strawberries growing within 50 meters of a strawberry field, more than 50 per cent of the wild plants contained marker genes from the cultivated strawberries".

A Danish team have shown a possible rapid spread of genes from oilseed rape to the weedy relative *Brassica campestris*.

There are other risks. *Brassica campestris*.

There are other risks. The introduction of Bt gene allows a drastic reduction in the use of toxic chemicals for crop protection. But a poorly controlled use of Bt-technology can destroy more effectively the predators than the pests.

Or when" many crop plants are transformed with similar effective traits, in such situation, many polyphagous pest species, which by nature are more flexible evolutionarily than those that have a narrower diet, are likely to overcome Bt resistance very quickly".

Therefore, before releasing a transgenic plant, any risk has to be weighed against the benefit of the transgenic crops. We must not forget that annually in the world 500.000 acute pesticide poisonings, with 5000 deaths, are observed.

DURING THE MICROPROPAGATION PROCESS, THE GENETIC STABILIT

During the micropropagation process, the genetic stability of the new shoots is dependent upon their origin. Axillary shoots issue from pre-existing buds and are normally true to type, indeed the meristematic cells are genetically very stable.

Adventitious shoots, such as somatic embryos, are neo-formed buds developed directly on some organs, or indirectly through a callus phase formed on this organ.

So, if the mother plant presents a cell mosaic or chimaeric tissues, risks of genetic variation exists. It is similar in the case of an indirect regeneration, when the callus phase is too long.

PROPAGATION BY AXILLARY SHOOTING

This technique has proved to be the most applicable and reliable method of *in vitro* propagation. Axillary shoot growth is stimulated by overcoming apical meristem dominance.

Commercial tissue culture laboratories are now able to propagate a large number of herbaceous ornamental species and several woody plants in this way. However, the propagation of Pelargonium, Howea, and a few other horticultural plants are always difficult to propagate by axillary branching.

Propagation by Direct or Indirect Organogenesis

Adventitious shoots could arise directly from the tissue of explants without callus formation. Several plants of the family gesneriaceae (Saintpaulia, Streptocarpus,...) regenerate directly buds on leaf explants, likewise Lilium rnerate directly buds on leaf explants, likewise Lilium regenerates on scales.

However more often, like for Ficus lyrata, adventitious buds appear on callus. While coffee, cocoa trees, and many conifers are produced by somatic embryogenesis developed on callus or cell suspensions.

Constraints of *in Vitro* Micropropagation

The establishment of axenic culture could be difficult when the explants are coming from hot and humid countries.

Moreover, in the course of micropropagation several authors observed a sudden appearance of endogeneous bacteria. Nevertheless only Holland and Pollaco (1994) have demonstrated the presence of this kind of bacteria in more than 70 different species. For Leifert and Woodward (1997), the major source of contamination is the initial explant. They add that microbial contamination in commercial plant tissue culture laboratories is the most important cause of losses. For this reason, they support the introduction of a microbiological production control strategy.

A second important constraint during the initial phase mainly concerns woody plants. A mature tree must be rejuvenated to recover its morphogenetic competence. Franclet (1981) described various treatments, micrografts, cascade grafts, cytokinin treatments to reverse the adult phase to juvenile. The micropropagation of Acacia Senegal was only possible after rejuvenation by micrografting (Palmnly possible after rejuvenation by micrografting.

A distinction between chronological, ontogenic and physiological age is needed to understand the rejuvenation concept. This was definitively demonstrated after a complete reversion of Sequoia sempervivens.

Improvement of Axillary Branching

The cost of micropropagated plantlets is also an important limitation of the techniques. In New Zealand, where they produce 2-3 million micropropagated radiata pine per annum, the relative cost of micropropagated planting stock had dropped from 13,8 times the cost of seedlings in 1988 to 6,9 by 1993.

To reduce manpower costs, several improvements have been proposed. The more simple method was *in vitro* layering developed by Wang (1977) to clone PVX-free potato plants. The first plantlets placed on the medium in a horizontal position developed axillary shoots. They are harvested by cutting one centimetre above the medium surface, at 3 weeks intervals. A similar technique called 'hedging system ' by Aitken Christie and Jones (1987) was later used to produce Pinus radiata. Ziv (1990) proposed for corm plants, Gladiolus and Nerine, a very rapid propagation system. She reduces the internodes and leaves by introduction of an anti-gibberellin agent in the medium. Finally, only aggregates of buds are formed, then they finally, only aggregates of buds are formed, then they are divided and introduced in bioreactors for mass production. Similar systems were developed in Gembloux, to propagate carub trees and asparagus. Since 1988, Duhem was producing very large quantities of Eucalyptus plantlets in Petri dishes without anti-gibberellin but in complete darkness. Transfers from one Petri to another is made by a simple squashing.

Somatic Embryogenesis Propagation

For genetically stable species, somatic embryogenesis offers a very fast scaling-up system, especially when it's possible to produce embryos in bioreactors. Unfortunately this production via fermentors was not as simple as first envisaged. And today, only a few model plants are successfully produced by such technology: carrot, celery. Other applications remain at the experimental stage: coffee, oil, palms, conifers, Euphorbia pulcherrima,... and several other horticultural species.

Several bottlenecks limit the use of this interesting technology. One of the main problems is genetic stability. So, despite the clonal nature of nucellar embryos, different morphological anomalies can occur among mango somatic embryos, as it was also observed in Citrus plants derived from nucellar cultures. Another difficulty is the loss of embryogenic capacity over time, a phenomenon observed with different species. It is also a phenomenon observed with different species. It is also important to note that somatic embryogenic lines of conifers are always originated from immature embryos.

Artificial Seeds

Another very interesting possibility of the somatic embryogenesis technology has been developed during the past fifteen years by Redenbaugh and his team (1991, 1993). They were able to encapsulate somatic embryos by hydrogel coatings (sodium alginate), producing single embryo artificial seeds. To date, some improvements offer the possibility to directly plant the artificial seeds in the greenhouse on special substrates (vermiculite, sans,...). This methodology will provide in the future a good technique to reduce the cost of transplants.

Gene Bank

Tissue culture methods offer the opportunity for *in vitro* collecting, rapid multiplication and distribution of important, elite, or rare plants that are threatened with extinction. The two major *in vitro* storage strategies are slow growth and cryopreservation. Since the first results of Seibert (1976), who was able to initiate shoots from carnation shoot apices frozen to -196C, this technique is now successful for many of horticultural species. Dereuddre *et al.* (1991) have provided a very simple technology to freeze encapsulated meristems in dried alginate beads. It works for pear, strawberry, eucalyptus, potato,....

Some Ior Pear, Strawberry, Eucalyptus, Potato

Some international institutions have very large collections of old and current varieties, available for exchange and introduction in crop improvement programmes. The International Potato Centre (CIP) in Lima, Peru, has a large world potato collection. Germplasm of sweet potato and cassava is at the International Institute of Tropical Agriculture (IITA), Ibadan, Nigeria.

The movement of germplasm involves the risks of accidentally introducing plant quarantine pests along with the host plant material. To limit these risks, the plant material should be transferred from one country to another as *in vitro* cultures through a transit Centre, where it should be indexed. For bananas, in the framework of INIBAP, the transit Centre is the Catholic University of Leuven in Belgium where a very large *in vitro* germplasm exists.

RESEARCH PREPAREDNESS FOR ACCELERATED GROWTH OF HORTICULTURE

The horticulture scenario of the country is rapidly changing. The production and productivity of horticultural crops have increased manifold. Production of fruits and vegetables has tripled in the last 50 years. The productivity has gone up by three times in banana and by 2.5 times in potato. Today horticultural crops cover about 25 per cent of total agricultural exports of the country. The corporate sector is also showing greater interest in horticulture. A major shift in consumption pattern of fresh and processed fruits and vegetables is expected in the coming century. There will be greater technology adoption both in traditional horticultural enterprise as well as in commercial horticulture sectors. Diversification and value addition will be the key words in the Indian horticulture in the 21st Century.

Horticulture research in India is about four decades old. Systematic research on fruit, vegetable and ornamental crops began in 1954 with the initiation of independent institutions and programmes. The research agenda is designed relevant to national plans and priorities for the horticulture development. Today, eight ICAR institutes with 27 regional stations, 1 project directorate, 10 national research centres, 16 all India coordinated research projects (AICRPS) with 223 research stations, 1 full-fledged university of horticulture, 25 state agricultural universities and 7 multi-disciplinary institutes of the ICAR are engaged in horticulture research. In addition, a few R&D establishments of crop/commodity boards and private sectors are providing research support to Indian horticulture. Research system in horticulture is now geared to provide necessary technological support to the expanding horticultural industry. The research efforts in the past were mainly concentrated on crop improvement, propagation of seed/planting material, agrotechniques, crop protection and post harvest management.

VARIETAL DEVELOPMENT

Among the fruit crops, improved high yielding mango varieties, Mallika, Amrapali, Ratna, Sindhu, Arka Aruna, Arka Puneet, Dashehari-51 and hybrids namely CISH-M-1 and CISH-M-2 have been developed. Mallika is coming up in southern states like Karnataka and Amrapali is performing well in eastern India. Dashehari-51, a regular bearing cultivar with about 38 per cent higher productivity than the normal Dashehari, has been identified after 14 years of rigourous selection. In guava, three selections, namely Lalit, CISH-G-1 and CISH-G-2 have been developed for domestic and export markets.

The fruits of Lalit are of medium size weighing about 150 g each and suitable for both table and processing purposes. In banana, high yielding hybrids like FHIA-01 and FHIA-03 are promising for replacing varieties, Panchananda and Bluggoe, respectively. Cultivar, Saba is found promising under sodic soils. In addition, high yielding varieties like Col, Hl and H2 have been developed. In grape, superior and high yielding varieties have been developed *e.g.* Beauty Seedless and Pusa Seedless and early ripening variety, Perlette for cultivation under North Indian conditions and Anab-e-Shahi, Dilkhush, Thompson Seedless, Tas-A-Ganesh, Sonaka, Bangalore Blue and Pachadraksha for south Indian conditions.

In citrus, high yielding and cluster bearing varieties of acid lime have been developed. *e.g.*, Rough Lemon, Rangpur Lime, Pramaini, Vikram PKM-1. Trifoliate oranges, namely Flying Dragon and Rich 16-6 are dwarfing types. In papaya, high yielding superior varieties both for table purpose and papain production have been developed. *e.g.*, Co-1 to Co-7, Coorg Honey Dew, Pusa Delicious, Pusa Majesty, Pusa Giant and Pusa Nanha. In apple, superior hybrids have been developed. *e.g.*, Lal Ambari, Sunehari. Red Spur, Star Crimson, Golden Spur, Red Chief, Oregon Spur, Skyline Supreme and Vance Delicious have been identified. Tissue culture protocols for micro-propagation of two commercial varieties have been developed.

Among the vegetable crops, more than 130 open pollinated varieties, 36 hybrids, 3 synthetics and 29 resistant varieties of 20 vegetable crops have been developed and released for cultivation in different agro-climatic regions. These include 40 in tomato, 45 in brinjal, 13 in cauliflower, 12 in chillies, 20 in pea, 9 in musk melon, 16 in onion and 44 in other crops.

In potato, 33 high yielding varieties have been developed indigenously for large scale cultivation in different regions. Kufri Ashoka and Kufri Pukhraj mature in 75 days. Kufri Jawahar and Kufri Satluj are field resistant to late blight. Kufri Jawahar has most ideal plant type for inter-cropping. Kufri

Chipsona-1 and Kufri Chipsona-2 have been developed with excellent processing attributes, comparable to exotic varieties. Kufri Swarna resistant to golden nematode is ideal for Nilgiri Hills.

In tuber crops, improved varieties of different tuber crops have been recommended/released for commercial cultivation. These includes 9 varieties of cassava, 15 varieties of sweet potato, 6 varieties of colocasia, 3 varieties each of greater yam and lesser yam, 1 each of *Amorphophallus,* taro and yambean. Cassava varieties, Sree Visakham and Sree Prakash are popular in Kerala. Triploid clone, Sree Harsha with high dry matter and starch content is suitable for industrial belt of Tamil Nadu. Two early maturing varieties Sree Jaya and Sree Vijya have been released for culinary purposes. Elephant foot yam variety, Am-15 has been released with high yield potential of 41 t/ha.

Among the plantation and spice crops, India is the first country to exploit hybrid vigour in coconut. Twelve hybrids involving tall and dwarf parents and 4 varieties have been released for commercial cultivation. These varieties yield 21 to 89 per cent more than the local cultivars. Some of the released varieties like Chandra Kalpa and Pratap (Banawali Green Round) are receiving wide acceptance of farmers. Chowghat Green Dwarf variety is good for tender coconut purpose. In arecanut, 4 high yielding varieties, namely, Mangala, Sumangala, Sreemangala and Mohitnagar have been developed, giving about 30 per cent higher yield than the local cultivars.

In oil palm, first efforts for improvement were made by producing Tenera hybrids using Pisifera pollen imported from Nigeria. Dura x Pisifera hybrids are field tested in East and West Godavari districts, Khammam and Krishna districts of Andhra Pradesh, with yield potential of 20-25 tonnes/ha of FFB from the fifth year. In cashew, 22 region specific selections and 12 hybrids with yield potential of 1.5-2 tonnes of raw nuts/ha have been produced and released for commercial cultivation.

The present standards fixed for cashew varieties include export grade kernels of W-210 to W-240 and at least one tonne per ha yield with 30 per cent shelling. In black pepper, 6 varieties, namely, Sreekara, Subhakara, Palode-2, Panniyur-2, Panniyur-4, Panchami and Pournami, and 2 hybrids *viz.* Panniyur-1 and Panniyur-4 have been developed.

In cardamom, a number of improved varieties have been developed and released for commercial cultivation *e.g.,* CCS-1, Mudigere-1, PV-1, ICRI-1 and ICRI-2. In ginger, varieties like Suprabha, Suruchi, Suravi and Varada have been developed. In turmeric, several varieties *viz.,* Co-1, Krishna, Sugandham, BSR-1, Suvarna, Roma, Suroma, Rajendra Sonia, Sugana, Sudarshana, Ranga, Rasmi, Prabha, Prathiba, Mega Turmeric and RCT-1 have been developed with yield potential of up to 44 tonnes of fresh rhizomes per ha. Three high yielding cinnamon lines, namely, Navashree, Nithyashree and Konkan Tej have been released for cultivation.

Agrotechniques

In fruit crops, improved agrotechniques developed have helped the farmers in improving the productivity and quality of produce. Soil application of paclobutrazol (4 g/tree) increase flowering and fruiting in mango on commercial scale in coastal Maharashtra. It also controls irregular bearing in cultivar Dashehari. Spray of NAA @ 200 ppm in October is recommended for control of malformation. Heavy fruit drop at maturity in cultivar Langra can be controlled by spraying NAA (20 ppm).

In guava, double spray of 10 and 20 per cent urea on cultivars, Allahabad Safeda and Sardar twice at bloom eliminate poor quality rainy season crop and increases winter season yield by 3 and 4 times, respectively. Application of neem coated urea (800 g/plants) yields 98 kg fruits/plant in variety, Sardar compared to 37 kg from untreated plants. In banana, high density planting (4550 plants/ha) yield up to 174 t/ha.

Adoption of improved technology in Maharashtra has resulted in fruit yield increase up to 52 t/ha. In citrus, two grafting methods using inverted 'T' cut and apical triangle cut have been developed with overall success of around 36 per cent using either method. For accelerating the survival of growth, shoot tip grafts, the successful grafts are double grafted (side grafted) on vigourous greenhouse grown Rough Lemon and Rangpur Lime seedlings. Rangpur Lime rootstock is found superior for sweet oranges and mandarins.

In papaya, closer spacing of 1.4 x 1.4 m is recommended for high yield. Drip irrigation techniques have been standardized. In banana, it has resulted in production gain (60-70 per cent) and early harvesting (40-50 days), besides improved water efficiency. Likewise, in grapes higher yields have been obtained with better water use efficiency (11 per cent).

In vegetable crops, improved production technology has been developed for major crops. Drip irrigation is economical in tomato and brinjal. In cucumber, replenishment of evaporation loss through irrigation resulted in maximization of yield of quality fruits. Drip irrigation in watermelon provided 33 per cent higher yield with a water saving of 40 per cent. Nutrient requirements and fertilizer schedules have been worked out crop-wise and recommended for different agro-climatic regions. In leguminous vegetables, high N depresses nodulation. The VAM fungi increases P availability to plants.

In all leguminous vegetables, inoculation of the VAM fungi along with *Rhizobium* culture is beneficial. Production technologies for *kharif* onion in northern India and long day type onions for high altitudes have been standardized. Pendimethalin (Stomp) has been found effective in controlling weeds in tomato, brinjal, chilli, bell pepper and okra. In potato, a number of potato-based multiple and inter cropping systems have been developed for different potato growing regions. Intercrop combinations with sugarcane in Maharashtra, wheat in Chhota Nagpur area and linseed in central Uttar Pradesh are found remunerative. A suitable method of urea application has also been worked out. In tuber crops, short duration legumes *viz.*, groundnut and French bean and cowpea can be successfully inter-cropped with cassava. Short-duration cassava, Sree Prakash is ideal in double cropped rice fields. Studies on cassava-based multiple cropping systems involving banana, coconut, *Leucaena* and Eucalyptus, have shown banana-cassava combination to give maximum root yield. Banana and coconut combination reduces soil loss and surface run-off considerably. *Dioscorea* and elephant foot yam with banana, Nendran can generate an additional income of ₹ 20,000/- over the sole crop of banana. For Inter-cropping *Dioscorea* in coconut garden, the ideal planting density is 9000 plants/ha. When *Amorphophallus* is raised as an inter-crop in coconut garden, one third dose of recommended fertilizers is sufficient. Inoculation with VAM fungi in cassava give about 15-20 per cent increase in yield.

Among the plantation and spice crops, density of 175 coconut palms/ha (7.5 m x 7.5 m spacing) is found ideal. In general, NPK application of 500:320:1200 g/palm/year is found optimum. A multi-storied cropping system involving black pepper trained on coconut trees, and cocoa in between the rows of coconut and pineapple in the ground floor has been found ideal for exploiting light, soil and air spaces. In arecanut, application of NPK (100:40:140 g) and green leaf (14 kg) per palm per year is recommended for coastal regions of Kerala and Karnataka and for plains of West Bengal, Karnataka and Assam.

In oil palm, application of NPK (1200:600:1200 g/palm) is found to give 17.1 tonne of FFB per ha. In black pepper, rapid methods for production of rooted cuttings of pepper have been developed and a commercial protocol has been standardised for micropropagation of black pepper. Ginger yield could be increased up to 33 per cent by application of neem cake at the rate of two tonnes per ha and the fertilizer schedule of 75 kg each of N, P_20_5 and K_20. Technology for storage of ginger seed rhizome is standardised and recommended.

Protection Technologies

For major fruit crops, plant protection schedules have been developed for the control of significant insect-pests for wider adoption. In the recent years, research efforts are directed to devise eco-friendly, economical and long lasting control measures. Success has been achieved in biological control of mealy bugs in mango and guava. The *Beauveria bassiana* has been found killing mango mealy bug and hopper. In grapes, integrated management of *Spodoptera* caterpillar involving light and pheromone traps, NPV and neem based insecticides and biological control of mealy bug by the beetle *Cryptolaemus montrouzieri* have been standardized. Studies on pesticide residues have resulted in working out of safe-waiting periods for harvesting and consumption of fruits.

In vegetable crops, about 50 improved measures for efficient management of diseases and 23 for insect-pests have been worked out and popularized in different agro-climatic regions in the country. Integrated pest management

(IPM) for controlling diamond back moth on cabbage through a trap crop like mustard has been demonstrated. Fruit borer *(H. armigera)* on tomato can be controlled by the release of *Trichogramma pretiosum* alone and in combination with HaNPV. In potato, integrated management schedules for control of bacterial wilt and tuber moth have been developed. A late blight forecasting system has been developed for the hills.

Among the tuber crops, the major diseases affecting tuber crops are cassava mosaic and brown leaf spot in cassava, *Phytophthora* leaf blight in colocasia, *Fusarium* wilt in elephant yam and virus diseases of sweet potato. Foliar sprays of Bavistin (0.1 per cent) combined with disodium and dipotassium phosphates (100 ppm) and calcium sulphate at 15-day interval was found to check foliar diseases in sweet potato. The major pests include spider mites, scale insects and white fly on cassava, weevil on sweet potato, defoliators, aphids and mites on colocasia, and scales and mealy bugs on yams and elephant yam. Cultural methods for weevil control include clean cultivation, destruction of alternate hosts and timely harvest. An effective IPM package using synthetic sex pheromone has been developed. Control measures involving insecticides have been evolved for the control of pests of other tuber crops.

Among the plantation and spice crops, bud rot of coconut caused by *Phytophthora palmivora* can be effectively controlled by spraying Bordeaux mixture. Calyxin root feeding and drenching of soil with 1 per cent Bordeaux mixture along with neem cake application @ 5 kg per palm per year is recommended for controlling Thanjavur wilt disease reported in Tamil Nadu, Andhra Pradesh and Karnataka.

A package of practices has been developed for managing mycoplasma like organisms (MLOs) in root wilt affected coconut palms in Kerala and Thatipaka disease affected palms in Andhra Pradesh.

Eradication of all root wilt affected palms is recommended. In cashew, tea mosquito bug (TMB) can be effectively controlled through a schedule of spray coinciding with flushing, flowering and fruiting. For effective control of stem and root borer infestation, constant monitoring and adoption of strict sanitation in the plantations coupled with prophylactic application of coal tar and kerosene in the ratio of 1:2 on trunks are recommended. In black pepper, spraying Bordeaux mixture (1 per cent) and drenching the soil with copper oxychloride (0.2 per cent) is found effective in managing *Phytophthora* foot rot.

Seed/planting Material Propagation

In most of the fruit crops, vegetative propagation techniques have been standardized. Soft wood grafting has been standardized for mango, sapota, custard-apple and jackfruit. Other vegetative propagation techniques have been developed for *ber, aonla,* jackfruit, custard-apple and *bael.* In mango, veneer grafting and stone-grafting is practised commercially. Mango variety, Vellaikolumban is suitable semi-dwarfing rootstock for Alphonso. Old unproductive mango trees can be rejuvenated successfully by pruning the 4th order branches during December-January. Flowering and fruiting are regular in pruned trees. For mandarin orange, Rangpur lime is a drought hardy rootstock. In grapes, Dogridge and Salt Creek (Ramsey) are suitable for minimizing adverse effects of soil salinity on Thompson Seedless.

A tissue culture technique for mass multiplication of Dogridge has been standardised. In banana, sword suckers of 700-1000 g are optimum. Rhizomes with active lateral buds and dead central buds are preferred for distant transportation in western India. Double paring and shade drying followed by dipping in Monocrotophos (0.5 per cent) and Bavistin (0.2 per cent) is recommended to disinfect nematodes and soil borne fungi. Tissue cultured banana plants are now commercially adopted for their uniformity in flowering and produce. Shoot-tip grafting technique in citrus has been considerably advanced. In vegetable crops, seed production of over 120 open pollinated high yielding varieties of different vegetables has been well established in the country. Hybrid seed production has become easier with the development of male sterile lines in tomato, self incompatible lines in cauliflower and gynoecious lines in cucumber and muskmelon. In brinjal, functional male sterility controlled by a single recessive gene has been identified.

Temperature barrier in cole crops (cabbage and cauliflower) has been overcome by developing heat tolerant hybrids. It is now possible to cultivate cabbage and cauliflower in southern India. Development of tomato varieties resistant to bacterial wilt has made their cultivation successful in non-traditional areas. Onion seed production technology for cultivation in kharif season has been developed for north Indian states especially, Haryana, Punjab and western Uttar Pradesh. Seed Plot Technique has been developed for production of disease-free potato seed in plains. It is widely adopted by farmers. A new technology for raising commercial crop of potato using 'True Potato Seed' (TPS) has been developed and standardized as supplementary technology to the traditional tuber grown crop. Two TPS populations, TPS-C-3 and HPS-113 are recommended for commercial production in Bihar, Gujarat, Tripura and West Bengal. Micropropagation protocols have been developed in banana, oil palm, cashew, black pepper, ginger, *etc*. Seed gardens of Tall (T) x Dwarf (D) and D x T hybrids have been established for production of coconut hybrids.

Post Harvest Management Technologies

The post harvest handling of fruits and vegetables accounts for 20-30 per cent of losses at different stages of storage, grading, packing, transport and finally at marketing as a fresh produce or in processed form. A number of improved technologies have been developed for commercial exploitation. An on-farm, low cost, environment friendly cool chamber, Zero Energy Cool Chamber has been developed using locally available material.

The principle of evaporative cooling reduces the inside temperature by as much as 17-18 °C and keeps the relative humidity above 90 per cent during peak summer. The chamber increases the shelf life and reduces PLW of banana, mango, orange lime, grape fruit, tomato and potato in different situations in India. Maturity standards for mango, guava, grape, litchi and *ber* and chemical treatments for regulation of ripening in mango, sapota and banana have been standardised.

Optimum storage temperatures worked out for several fruits, vegetables and tuber crops. A mango harvester, fruit peeler, hand and pedal operator cassava chipping machines, harvesting tools (5-14 times efficient), coconut dehusking machine, implements for mechanization of potato cultivation and other crops have been developed. A number of improved technologies have been developed for commercial exploitation *viz.*, tent type foldable solar dryer, packaging boxes for distant transportation of apple, mango, citrus and plum, production of value added products-pectin from peel and flour from mango fruit kernel, production of fruit post carbonated beverages *etc.*

Production Constraints

In spite of great strides made, the productivity of horticultural crops, in general, is still quite low and the post harvest losses particularly of perishable commodities, are considerable. Improvement in quality standards of the produce and their marketing are essential to increase our share in the global market. The research agenda in horticulture is by design relevant to national plans and priorities and research programmes are normally formulated keeping in view the thrust areas in development. The major technology related constraints contributing to low productivity of horticultural crops and inferior quality of produce are:

- Vast majority of holdings are small and un-irrigated.
- Large tracts of low and unproductive plantations needing replacement/ rejuvenation.
- Low productivity of crops due to inferior genetic stocks and poor management.
- Inadequate supply of quality planting materials of improved varieties.
- High incidence of pests and diseases.
- Heavy post harvest losses and low utilization in processing sector.

For addressing the above constraints, research institutions are engaged in both basic and applied research. While formulating research strategies some of the inherent weaknesses associated with perennial tree crops and certain perpetual problems in Indian horticulture must be kept in mind. They are:

- Long period required for development of improved genotypes. Application of biotechnological tools/methods in horticultural crops is still in its early stage of development in the country.

- Chronic production problems due to major disorders like alternate bearing, malformation and spongy tissue in mango, guava wilt, citrus decline, root wilt in coconut, viral disease in vegetables, *Phytophthora* diseases in large number of crops *etc*. remained largely unresolved.
- Lack of advanced technologies for post harvest handling, processing and marketing of produce.

Losses caused by biotic stresses are very high and due to pesticide residue problems development of eco-friendly IPM strategy is more relevant in horticulture. There is a threat for loss of valuable genetic resources, if measures are not taken for their conservation.

Wastelands and hilly terrains being the potential future expansion areas, matching technologies for dry land and hill horticulture need to be developed. Counter seasonal advantages from diverse agro-climatic situations provide strength for extended availability of horticultural crops round the year and such potentials can be harnessed only with relevant research support.

PLANT TISSUE SYSTEMS

STRUCTURE AND FUNCTION

Although the architecture varies considerably among roots, stems, and leaves, all plant organs are comprised of 3 basic tissue types: dermal tissue, vascular tissue, and ground tissue. A tissue is simply an assemblage of cells that work together as a structural or functional unit. In this lab, we will examine the general arrangements of those tissues among different plant organs. You will find that the distribution of tissues varies somewhat among different plants as well as among different organ systems.

For example, one of the key differences between monocots and dicots has to do with the arrangement of vascular tissues in leaves and stems. In next week's lab, we will continue our examination of plant tissues by exploring some of the different cell types that form these three plant tissue sytems.

DERMAL TISSUE

The dermal tissue system is derived from the protoderm and consists of 2 tissue types*:

- *Periderm*: Periderm includes cork and cork-producing tissues and is only found in plants with secondary growth.
- *Epidermis*: The epidermis constitutes the outmost layer of cells of roots, stems and leaves. Epidermal cells are usually covered by a waxy layer, called a cuticle that serves for waterproofing the cells.

VASCULAR TISSUE

Vascular tissues, derived from the procambium*, are the tissues responsible for transporting substances from one point in the plant body to another. As is the case with animals, the vascular tissues are responsible for the movement of dissolved gases, sugars, and minerals throughout the organism.

In plants, there are 2 types of conducting tissues:

- *Xylem*: Xylem is the tissue specialized for the conduction water and dissolved minerals. Cells of the xylem are dead at maturity and act as the structural medium for the movement of water. Individual xylem cells have multiple holes in the walls to allow for the passage of water between adjacent cells. The bulk of wood consists of xylem tissue.
- *Phloem*: Phloem conducts dissolved sugars throughout the plant. The main conducting cells of phloem, called"sieve elements", are alive at maturity, but are highly modified for carrying out the function of transport. For example, many of the organelles you might expect to find in a living plant cell, such as nuclei, ribosomes, Golgi apparati, cytoskeleton, and vacuoles are absent from the sieve elements in order to make more space for fluid transport. The small amount of protoplast in sieve elements is distributed in a thin layer along the periphery of the cell wall. In some cases, you may be able to observe smaller cells called"companion cells" distributed among the sieve elements. Companion cells contain all the organelles you would expect to find in a typical plant cell and carry out many of the necessary life functions for the sieve elements which lack organelles.

GROUND TISSUE

Derived from the ground meristem*, ground tissues are distributed between and among the dermal and vascular tissues of roots, stems, and leaves.

Ground tissues include a wide variety of tissue types that perform a multitude of functions within the plant.

Some of the more familiar ground tissues include the following:

- *Cortex*: the ground tissue distributed between the epidermis and vascular tissues in roots and stems. Cortex is usually modified for storage, especially of starches. In herbaceous green stems, cortex often contains chloroplasts and thus is further modified for carrying out photosynthesis.
- *Pith*: like cortex, pith is generally modified for starch storage, but is only found to the interior of the vascular tissues in stems and roots.
- *Mesophyll*: the ground tissue of leaves found between the two layers of epidermis. Leaf mesophyll is the tissue most adapted for photosynthesis and thus is generally rich with chloroplasts.

In addition to the functions indicated above, ground tissues also have an important role in wound healing and the regeneration of damaged or dead cells throughout the plant.

PLANT TISSUES ARE DERIVED FROM MERISTEMS

Unlike animals, most plants grow (increase the number and size of their cells) throughout their lifetime. This ability to grow indeterminately is due to a special form of tissue known as a"meristem".

A meristem is a region of tissue that is perpetually embryonic - that is, it is relatively undifferentiated and divides continuously, giving rise to all new cells and tissues that make up the plant body. Can you think of an analogy to the plant's meristem in animals/humans?

There are several meristems distributed throughout the plant body which are named for their location and/or for the tissue systems that are derived from them.

APICAL (PRIMARY) MERISTEMS

Are located in the tips (or, apex) of the growing roots and shoots of all plants. Primary meristems produce primary tissues and account for all the primary growth of plants. Primary growth is growth that contributes to elongation of the plant (height and root lengthening) and for the establishment of its basic organ structure (roots, stems, leaves).

LATERAL (SECONDARY) MERISTEMS

Lateral (Secondary) Meristems are organized as concentric cylinders (think of a can inside a can inside a can) within the plant body and are responsible for secondary growth. Secondary growth contributes to the thickening of a plant structure, or increases in girth. With few exceptions, lateral meristems tend to be present only in woody plants, such as trees, shrubs, and some vines.

As cells in the apical meristem divide and give rise to new cells, the tips of roots and shoots extend outward, away from the central plant body. As a consequence of this elongation due to the addition of new cells, the slightly older meristematic cells become distributed into different regions within the root and shoot. These slightly more mature cells still retain their meristematic properties, but are somewhat differentiated from the true apical meristem. These regions of meristematic tissue derived from the apical meristem are referred to as"primary meristems" and are named for the tissues that they will give rise to:

- *Protoderm*: Gives rise to dermal tissues,
- *Ground Meristem*: Gives rise to ground tissues, and
- *Procambium*: Gives rise to vascular tissue (aka,"cambial tissue").

The organization of primary meristems also varies among plant types and organs.

Like other organisms, the cells in a plant are grouped together into various tissues. These tissues can be simple, consisting of a single cell type, or complex, consisting of more than one cell type.

DERMAL TISSUE SYSTEM

The dermal tissue system consists of the epidermis and the periderm. The epidermis is a single layer of closely packed cells. It both covers and protects the plant. It can be thought of as the plant's"skin." Depending on the part of the plant that it covers, the dermal tissue system can be specialized to a certain extent. For instance, the epidermis of a plant's leaves secretes a coating called the cuticle that helps the plant retain water.

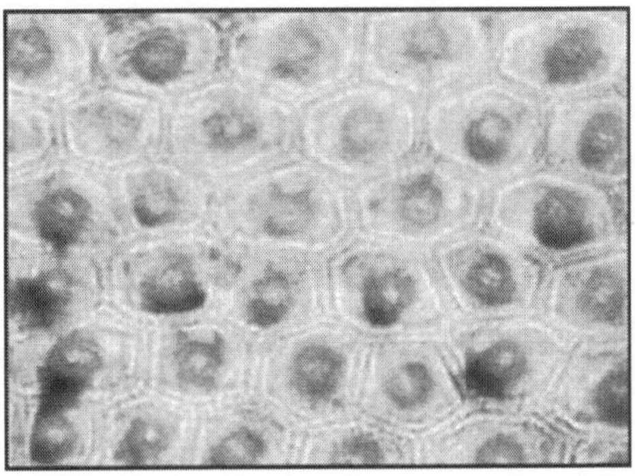

EPIDERMIS CELLS

The periderm, also called bark, replaces the epidermis in plants that undergo secondary growth. The periderm consists of cork cells and protects the plant from pathogens, prevents excessive water loss and provides insulation for the plant. Xylem and phloem throughout the plant make up the vascular tissue system. It allows water and other nutrients to be transported throughout the plant.

GROUND TISSUE SYSTEM

The ground tissue system synthesizes organic compounds, supports the plant and provides storage for the plant. It is mostly made up of parenchyma cells but can also include some collenchyma and sclerenchyma cells as well.

PLANT TISSUE CULTURE

Practically any plant transformation experiment relies at some point on tissue culture. There are some exceptions to this generalization, but the ability to regenerate plants from isolated cells or tissues *in vitro* underpins most plant transformation systems.

PLASTICITY AND TOTIPOTENCY

Two concepts, plasticity and totipotency, are central to understanding plant cell culture and regeneration. Plants, due to their long life span, have developed a greater ability to endure extreme conditions and predation than have animals. Many of the processes involved in plant growth and development adapt to environmental conditions. This plasticity allows plants to alter their metabolism, growth and development to best suit their environment.

Particularly important aspects of this adaptation, as far as plant tissue culture and regeneration are concerned, are the abilities to initiate cell division from almost any tissue of the plant and to regenerate lost organs or undergo different developmental pathways in response to particular stimuli. When plant cells and tissues are cultured *in vitro* they generally exhibit a very high degree of plasticity, which allows one type of tissue or organ to be initiated from another type. In this way, whole plants can be subsequently regenerated. This regeneration of whole organisms depends upon the concept that all plant cells can, given the correct stimuli, express the total genetic potential of the parent plant. This maintenance of genetic potential is called 'totipotency'. Plant cell culture and regeneration do, in fact, provide the most compelling evidence for totipotency.

The Culture Environment

When cultured *in vitro*, all the needs, both chemical and physical, of the plant cells have to met by the culture vessel, the growth medium and the external environment (light, temperature, etc.). The growth medium has to supply all the essential mineral ions required for growth and development. In many cases (as the biosynthetic capability of cells cultured *in vitro* may not replicate that of the parent plant), it must also supply additional organic supplements such as amino acids and vitamins.

Many plant cell cultures, as they are not photosynthetic, also require the addition of a fixed carbon source in the form of a sugar (most often sucrose). One other vital component that must also be supplied is water, the principal biological solvent. Physical factors, such as temperature, pH, the gaseous environment, light (quality and duration) and osmotic pressure, also have to be maintained within acceptable limits.

Plant Cell Culture Media

Culture media used for the *in vitro* cultivation of plant cells are composed of three basic components:

- Essential elements, or mineral ions, supplied as a complex mixture of salts;
- An organic supplement supplying vitamins and/or amino acids; and
- A source of fixed carbon; usually supplied as the sugar sucrose.

For practical purposes, the essential elements are further divided into the following categories:

- Macroelements (or macronutrients);
- Microelements (or micronutrients); and
- An iron source.

Media Components

It is useful to briefly consider some of the individual components of the stock solutions.

Macroelements

As is implied by the name, the stock solution supplies those elements required in large amounts for plant growth and development. Nitrogen, phosphorus, potassium, magnesium, calcium and sulphur (and carbon, which is added separately) are usually regarded as macroelements. These elements usually comprise at least 0.1 per cent of the dry weight of plants.

Nitrogen is most commonly supplied as a mixture of nitrate ions (from the KNO_3) and ammonium ions (from the NH_4NO_3). Theoretically, there is an advantage in supplying nitrogen in the form of ammonium ions, as nitrogen must be in the reduced form to be incorporated into macromolecules. Nitrate ions therefore need to be reduced before incorporation. However, at high concentrations, ammonium ions can be toxic to plant cell cultures and uptake of ammonium ions from the medium causes acidification of the medium.

In order to use ammonium ions as the sole nitrogen source, the medium needs to be buffered. High concentrations of ammonium ions can also cause culture problems by increasing the frequency of vitrification (the culture appears pale and 'glassy' and is usually unsuitable for further culture). Using a mixture of nitrate and ammonium ions has the advantage of weakly buffering the medium as the uptake of nitrate ions causes OH^- ions to be excreted.

Phosphorus is usually supplied as the phosphate ion of ammonium, sodium or potassium salts. High concentrations of phosphate can lead to the precipitation of medium elements as insoluble phosphates.

Microelements

These elements are required in trace amounts for plant growth and development, and have many and diverse roles. Manganese, iodine, copper, cobalt, boron, molybdenum, iron and zinc usually comprise the microelements, although other elements such as nickel and aluminium are frequently found in some formulations.

Iron is usually added as iron sulphate, although iron citrate can also be used. Ethylene Diamine Tetra acetic Acid (EDTA) is usually used in conjunction with the iron sulphate. The EDTA complexes with the iron so as to allow the slow and continuous release of iron into the medium. Uncomplexed iron can precipitate out of the medium as ferric oxide.

Organic Supplements

Only two vitamins, thiamine (vitamin B1) and myo-inositol (considered a B vitamin) are considered essential for the culture of plant cells *in vitro*.

However, other vitamins are also added to plant cell culture media. Amino acids are also commonly included in the organic supplement. The most frequently used is glycine (arginine, asparagine, aspartic acid, alanine, glutamic acid, glutamine and proline are also used), but in many cases its inclusion is not essential. Amino acids provide a source of reduced nitrogen and, like ammonium ions, uptake causes acidification of the medium. Casein hydrolysate can be used as a relatively cheap source of a mix of amino acids.

Carbon Source

Sucrose is cheap, easily available, readily assimilated and relatively stable and is therefore the most commonly used carbon source. Other carbohydrates (such as glucose, maltose, galactose and sorbitol) can also be used, and in specialized circumstances may prove superior to sucrose.

Gelling Agents

Media for plant cell culture *in vitro* can be used in either liquid or 'solid' forms, depending on the type of culture being grown. For any culture types that require the plant cells or tissues to be grown on the surface of the medium, it must be solidified (more correctly termed 'gelled'). Agar, produced from seaweed, is the most common type of gelling agent, and is ideal for routine applications. However, because it is a natural product, the agar quality can vary from supplier to supplier and from batch to batch. For more demanding applications, a range of purer (and in some cases, considerably more expensive) gelling agents are available.

These components, then, are the basic 'chemical' necessities for plant cell culture media. However, other additions are made in order to manipulate the pattern of growth and development of the plant cell culture.

PLANT GROWTH REGULATORS

We have already briefly considered the concepts of plasticity and totipotency. The essential point as far as plant cell culture is concerned is that, due to this plasticity and totipotency, specific media manipulations can be used to direct the development of plant cells in culture.

Plant growth regulators are the critical media components in determining the developmental pathway of the plant cells. The plant growth regulators used most commonly are plant hormones or their synthetic analogues.

Classes of Plant Growth Regulators

There are five main classes of plant growth regulator used in plant cell culture, namely:

1. Auxins;
2. Cytokinins;

3. Gibberellins;
4. Abscisic acid;
5. Ethylene.

Each class of plant growth regulator will be briefly looked at.

Auxins

Auxins promote both cell division and cell growth The most important naturally occurring auxin is IAA (indole-3-acetic acid), but its use in plant cell culture media is limited because it is unstable to both heat and light. Occasionally, amino acid conjugates of IAA (such as indole-acetyl-L-alanine and indole-acetyl-L-glycine), which are more stable, are used to partially alleviate the problems associated with the use of IAA. It is more common, though, to use stable chemical analogues of IAA as a source of auxin in plant cell culture media. 2,4-Dichlorophenoxyacetic acid (2,4-D) is the most commonly used auxin and is extremely effective in most circumstances. Other auxins are available, and some may be more effective or 'potent' than 2,4-D in some instances.

CYTOKININS

Cytokinins promote cell division. Naturally occurring cytokinins are a large group of structurally related (they are purine derivatives) compounds. Of the naturally occurring cytokinins, two have some use in plant tissue culture media. These are zeatin and 2iP (2-isopentyl adenine). Their use is not widespread as they are expensive (particularly zeatin) and relatively unstable. The synthetic analogues, kinetin and BAP (benzylaminopurine), are therefore used more frequently. Non-purine-based chemicals, such as substituted phenylureas, are also used as cytokinins in plant cell culture media. These substituted phenylureas can also substitute for auxin in some culture systems.

Gibberellins

There are numerous, naturally occurring, structurally related compounds termed 'gibberellins'. They are involved in regulating cell elongation, and are agronomically important in determining plant height and fruit-set. Only a few of the gibberellins are used in plant tissue culture media, Gibberelic Acid 3 (GA3) being the most common.

Abscisic Acid

Abscisic acid (ABA) inhibits cell division. It is most commonly used in plant tissue culture to promote distinct developmental pathways such as somatic embryogenesis.

Ethylene

Ethylene is a gaseous, naturally occurring, plant growth regulator most commonly associated with controlling fruit ripening, and its use in plant tissue culture is not widespread. It does, though, present a particular problem for plant tissue culture.

Some plant cell cultures produce ethylene, which, if it builds up sufficiently, can inhibit the growth and development of the culture. The type of culture vessel used and its means of closure affect the gaseous exchange between the culture vessel and the outside atmosphere and thus the levels of ethylene present in the culture.

Plant Growth Regulators and Tissue Culture

Generalisations about plant growth regulators and their use in plant cell culture media have been developed from initial observations made in the 1950s. There is, however, some considerable difficulty in predicting the effects of plant growth regulators: this is because of the great differences in culture response between species, cultivars and even plants of the same cultivar grown under different conditions.

However, some principles do hold true and have become the paradigm on which most plant tissue culture regimes are based. Auxins and cytokinins are the most widely used plant growth regulators in plant tissue culture and are usually used together, the ratio of the auxin to the cytokinin determining the type of culture established or regenerated. A high auxin to cytokinin ratio generally favours root formation, whereas a high cytokinin to auxin ratio favours shoot formation. An intermediate ratio favours callus production.

CULTURE TYPES

Cultures are generally initiated from sterile pieces of a whole plant. These pieces are termed 'explants', and may consist of pieces of organs, such as leaves or roots, or may be specific cell types, such as pollen or endosperm. Many features of the explant are known to affect the efficiency of culture initiation. Generally, younger, more rapidly growing tissue (or tissue at an early stage of development) is most effective. Several different culture types most commonly used in plant transformation studies will now be examined in bit detail.

CALLUS

Explants, when cultured on the appropriate medium, usually with both an auxin and a cytokinin, can give rise to an unorganised, growing and dividing mass of cells. It is thought that any plant tissue can be used as an explant, if the correct conditions are found. In culture, this proliferation can be maintained more or less indefinitely, provided that the callus is subcultured on to fresh medium periodically.

During callus formation there is some degree of dedifferentiation (*i.e.* the changes that occur during development and specialization are, to some extent, reversed), both in morphology (callus is usually composed of unspecialised parenchyma cells) and metabolism. One major consequence of this dedifferentiation is that most plant cultures lose the ability to photosynthesise. This has important consequences for the culture of callus tissue, as the metabolic profile will probably not match that of the donor plant. This necessitates the addition of other components—such as vitamins and, most importantly, a carbon source—to the culture medium, in addition to the usual mineral nutrients.

Callus culture is often performed in the dark (the lack of photosynthetic capability being no drawback) as light can encourage differentiation of the callus. During long-term culture, the culture may lose the requirement for auxin and/ or cytokinin. This process, known as 'habituation', is common in callus cultures from some plant species (such as sugar beet). Callus cultures are extremely important in plant biotechnology. Manipulation of the auxin to cytokinin ratio in the medium can lead to the development of shoots, roots or somatic embryos from which whole plants can subsequently be produced. Callus cultures can also be used to initiate cell suspensions, which are used in a variety of ways in plant transformation studies.

CELL-SUSPENSION CULTURES

Callus cultures, broadly speaking, fall into one of two categories: compact or friable. In compact callus the cells are densely aggregated, whereas in friable callus the cells are only loosely associated with each other and the callus becomes soft and breaks apart easily. Friable callus provides the inoculum to form cell-suspension cultures. Explants from some plant species or particular cell types tend not to form friable callus, making cell-suspension initiation a difficult task.

The friability of callus can sometimes be improved by manipulating the medium components or by repeated subculturing. The friability of the callus can also sometimes be improved by culturing it on 'semi-solid' medium (medium with a low concentration of gelling agent). When friable callus is placed into a liquid medium (usually the same composition as the solid medium used for the callus culture) and then agitated, single cells and/or small clumps of cells are released into the medium. Under the correct conditions, these released cells continue to grow and divide, eventually producing a cell-suspension culture. A relatively large inoculum should be used when initiating cell suspensions so that the released cell numbers build up quickly.

The inoculum should not be too large though, as toxic products released from damaged or stressed cells can build up to lethal levels. Large cell clumps can be removed during subculture of the cell suspension. Cell suspensions can be maintained relatively simply as batch cultures in conical flasks. They are continually cultured by repeated subculturing into fresh medium. This results in dilution of the suspension and the initiation of another batch growth cycle. The degree of dilution during subculture should be determined empirically for each culture. Too great a degree of dilution will result in a greatly extended lag period or, in extreme cases, death of the transferred cells.

After subculture, the cells divide and the biomass of the culture increases in a characteristic fashion, until nutrients in the medium are exhausted and/or toxic by-products build up to inhibitory levels—this is called the 'stationary phase'. If cells are left in the stationary phase for too long, they will die and the culture will be lost. Therefore, cells should be transferred as they enter the stationary phase. It is therefore important that the batch growth-cycle parameters are determined for each cell-suspension culture.

Protoplasts

Protoplasts are plant cells with the cell wall removed. Protoplasts are most commonly isolated from either leaf mesophyll cells or cell suspensions, although other sources can be used to advantage. Two general approaches to removing the cell wall (a difficult task without damaging the protoplast) can be taken—mechanical or enzymatic isolation.

Mechanical isolation, although possible, often results in low yields, poor quality and poor performance in culture due to substances released from damaged cells. Enzymatic isolation is usually carried out in a simple salt solution with a high osmoticum, plus the cell wall degrading enzymes. It is usual to use a mix of both cellulase and pectinase enzymes, which must be of high quality and purity. Protoplasts are fragile and easily damaged, and therefore must be cultured carefully.

Liquid medium is not agitated and a high osmotic potential is maintained, at least in the initial stages. The liquid medium must be shallow enough to allow aeration in the absence of agitation. Protoplasts can be plated out on to solid medium and callus produced. Whole plants can be regenerated by organogenesis or somatic embryogenesis from this callus. Protoplasts are ideal targets for transformation by a variety of means.

ROOT CULTURES

Root cultures can be established *in vitro* from explants of the root tip of either primary or lateral roots and can be cultured on fairly simple media. The growth of roots *in vitro* is potentially unlimited, as roots are indeterminate organs. Although the establishment of root cultures was one of the first achievements of modern plant tissue culture, they are not widely used in plant transformation studies.

SHOOT TIP AND MERISTEM CULTURE

The tips of shoots (which contain the shoot apical meristem) can be cultured *in vitro*, producing clumps of shoots from either axillary or adventitious buds. This method can be used for clonal propagation. Shoot meristem cultures are potential alternatives to the more commonly used methods for cereal regeneration as they are less genotype-dependent and more efficient (seedlings can be used as donor material).

Embryo Culture

Embryos can be used as explants to generate callus cultures or somatic embryos. Both immature and mature embryos can be used as explants. Immature, embryo-derived embryogenic callus is the most popular method of monocot plant regeneration.

Microspore Culture

Haploid tissue can be cultured *in vitro* by using pollen or anthers as an explant. Pollen contains the male gametophyte, which is termed the 'microspore'. Both callus and embryos can be produced from pollen. Two main approaches can be taken to produce *in vitro* cultures from haploid tissue.

The first method depends on using the anther as the explant. Anthers (somatic tissue that surrounds and contains the pollen) can be cultured on solid medium (agar should not be used to solidify the medium as it containsinhibitory substances). Pollen-derived embryos are subsequently produced via dehiscence of the mature anthers. The dehiscence of the anther depends both on its isolation at the correct stage and on the correct culture conditions. In some species, the reliance on natural dehiscence can be circumvented by cutting the wall of the anther, although this does, of course, take a considerable amount of time. Anthers can also be cultured in liquid medium, and pollen released from the anthers can be induced to form embryos, although the effi-ciency of plant regeneration is often very low. Immature pollen can also be extracted from developing anthers and cultured directly, although this is a very time-consuming process.

Both methods have advantages and disadvantages. Some beneficial effects to the culture are observed when anthers are used as the explant material. There is, however, the danger that some of the embryos produced from anther culture will originate from the somatic anther tissue rather than the haploid microspore cells. If isolated pollen is used there is no danger of mixed embryo formation, but the efficiency is low and the process is time-consuming. In microspore culture, the condition of the donor plant is of critical importance, as is the timing of isolation. Pretreatments, such as a cold treatment, are often found to increase the efficiency. These pretreatments can be applied before culture, or, in some species, after placing the anthers in culture.

Plant species can be divided into two groups, depending on whether they require the addition of plant growth regulators to the medium for pollen/anther culture; those that do also often require organic supplements, *e.g.* amino acids. Many of the cereals (rice, wheat, barley and maize) require medium supplemented with plant growth regulators for pollen/anther culture.

Regeneration from microspore explants can be obtained by direct embryogenesis, or via a callus stage and subsequent embryogenesis. Haploid tissue cultures can also be initiated from the female gametophyte (the ovule). In some cases, this is a more efficient method than using pollen or anthers.

The ploidy of the plants obtained from haploid cultures may not be haploid. This can be a consequence of chromosome doubling during the culture period. Chromosome doubling (which often has to be induced by treatment with chemicals such as colchicine) may be an advantage, as in many cases haploid plants are not the desired outcome of regeneration from haploid tissues. Such plants are often referred to as 'di-haploids', because they contain two copies of the same haploid genome.

PLANT REGENERATION

Having looked at the main types of plant culture that can be established *in vitro*, we can now look at how whole plants can be regenerated from these cultures.In broad terms, two methods of plant regeneration are widely used in plant transformation studies, *i.e.* somatic embryogenesis and organogenesis.

SOMATIC EMBRYOGENESIS

In somatic (asexual) embryogenesis, embryo-like structures, which can develop into whole plants in a way analogous to zygotic embryos, are formed from somatic tissues. These somatic embryos can be produced either directly or indirectly. In direct somatic embryogenesis, the embryo is formed directly from a cell or small group of cells without the production of an intervening callus. Though common from some tissues (usually reproductive tissues such as the nucellus, styles or pollen), direct somatic embryogenesis is generally rare in comparison with indirect somatic embryogenesis.

In indirect somatic embryogenesis, callus is first produced from the explant. Embryos can then be produced from the callus tissue or from a cell suspension produced from that callus. Somatic embryogenesis usually proceeds in two distinct stages. In the initial stage (embryo initiation), a high concentration of 2,4-D is used. In the second stage (embryo production) embryos are produced in a medium with no or very low levels of 2,4-D.

In many systems it has been found that somatic embryogenesis is improved by supplying a source of reduced nitrogen, such as specific amino acids or casein hydrolysate.

Organogenesis

Somatic embryogenesis relies on plant regeneration through a process analogous to zygotic embryo germination. Organogenesis relies on the production of organs, either directly from an explant or from a callus culture. There are three methods of plant regeneration via organogenesis. The first two methods depend on adventitious organs arising either from a callus culture or directly from an explant. Alternatively, axillary bud formation and growth can also be used to regenerate whole plants from some types of tissue culture. Organogenesis relies on the inherent plasticity of plant tissues, and is regulated by altering the components of the medium. In particular, it is the auxin to cytokinin ratio of the medium that determines which developmental pathway the regenerating tissue will take. It is usual to induce shoot formation by increasing the cytokinin to auxin ratio of the culture medium. These shoots can then be rooted relatively simply.

PLANT TISSUE ANALYSIS

Crop nutrient uptake is influenced by many factors other than the soil test. Soil testing and plant analysis are designed to work together. Soil testing identifies the soils nutrient reserves and predicts the nutrient needs, while plant analysis identifies the actual nutrient uptake.

Plant tissue analysis is a useful diagnostic tool and can greatly help with fertility management. From emergence through the first six weeks of growth, plant analysis is helpful in identifying nutrient uptake. Testing corn leaf samples at this early stage may indicate where additional nitrogen should be applied. This is a way to determine the cost-effectiveness of the additional application of fertilizer. Plant analysis may be utilized at the reproductive stage (full grown) to determine nutrient uptake for that season. This allows for determination of fertility requirements for the upcoming growing season. Tissue sampling and analysis is a key tool in determining the soil's ability to meet the sufficient nutrient requirements of the crop, thereby reducing any possible nutrient stress that may result in a potential yield drag or loss. A variety of tests are available, including phosphorus, potassium, calcium, magnesium, sulfur, sodium, zinc, iron, manganese, copper, nitrogen, and boron.

Remember: test results are only as good as the sample. The most common mistake in plant sampling is not obtaining enough tissue for an accurate analysis. Several leaves from several plants will ensure a more representative sampling and ensure enough plant material. Fertility management samples should be collected from representative areas of the field. Always take composite samples. One plant is not a sample. Generally, select only upper mature leaves for your tissue sample. When samples are being collected for nutrient review, specific leaves are normally selected depending upon the plant. Here are a few examples:

- Corn-at tasseling, collect the leaf at the ear node
- Soybeans-at flowering collect the most mature trifoliant from the top of the plant
- Wheat and small grain-at beginning of head set collect the flag leaf
- Alfalfa-at flowering collect the top 6 inches of the plant

When troubleshooting is the issue, it is extremely important to take leaves from several plants experiencing the same problem and use this composite for one sample. Compare this analysis with a composite sample taken from the non-affected area of the field. Sample #2 is the control sample. This gives the grower an idea of the sufficient nutrient level for that crop in that field at that sampling period.

For example: a portion of a corn field is exhibiting chlorosis (yellowing) and stunted growth. Plant tissue analysis is the best way to determine the reason for the chlorosis and abnormal growth. A composite sample of upper mature leaves from 6 to 10 chlorotic plants and a composite sample taken from the upper mature leaves from 6 to 10 plants located in the non-affected area of the field are submitted for analysis.

Where to Sample

Avoid leaves that were heavily preyed upon by insects or mechanically damaged. Avoid taking samples from areas located in border rows or heavy shade. Do not collect tissue that is already dead. Do not take samples from contaminated or chemically altered plants. Take diagnostic samples from the affected site for one composite sample and one from the non-affected area for a second composite sample. If the entire field is distressed, collect a composite sample over the entire area and note that the entire field is affected.

When to Sample

Timing of diagnostic sampling depends on the first sign of trouble. The earlier the problem is detected, the easier it is to treat with fertilizer or lime application. If nutrient application is not feasible due to plant growth, these results, along with soil analysis, can be used in fertilizer recommendations for the following growing season. Fertility management sampling is done at the time of flowering or pollination (reproductive stage). Avoid taking samples too late in the growing season. Take a single composite sample at the onset of the reproductive stage to measure nutrient uptake throughout the growing season. Take samples when the plant is still viable and growing. It is too late once it has started to produce seed. These values determine the following growing season's fertility needs.

SUBMITTING THE SAMPLE

Place each dried composite sample in a paper bag. Do not use plastic bags for plant samples because the plants mold. Label each bag with the field name and sample number. Please label them good, bad, affected or non-affected. Please include information that contains: name, address, telephone number, field names, number of samples per field and the type of test.

A plant's ability to take up adequate nutrients is influenced by many factors such as soil fertility level, soil temperature, soil compaction, soil moisture, nutrient balance in the soil, fertilizer applications, plant genetics, and many other factors. A chemical soil test cannot accurately account for all of these factors. A producer must use plant analysis if he expects to identify and understand the nutrient efficiency and needs of his crops and soils. Only plant analysis can tell us the actual nutrient status of the crop. Plant analysis can be used to detect low nutrient levels in plants before they are seen by the human eye. Once visual symptoms are visible, top yields and quality have already been compromised. Plant analysis is an important part of planning future fertility programmes.

Sampling Guide

Our plant sampling guide contains information on most plant species. If you need furthur assistance, contact the lab for specific information on your crop.

USES OF PLANT ANALYSIS

- Plant analysis can be used by turf managers to:
- Confirm suspected nutrient deficiency symptoms;
- Verify toxicities;
- Reveal hidden hunger (*i.e.* plants show no visible symptoms, but the nutrient content is low enough to reduce growth or affect quality characteristics);
- Aid in evaluating the efficiency of applied fertilisers;
- Assist in formulating fertilisation practices; and,
- Monitor the nutrient status of plants throughout the growing season to determine whether each nutrient is present in sufficient concentration for optimum growth characteristics.

Plant analysis is a proven and effective means of predicting fertilizer needs for many plant species. However, it does not completely replace a soil test. Soil and plant analysis serve different purposes and when properly used they compliment each other in providing detailed information for maximizing the efficiency of fertility programmes. Soil testing is based on the concept that the concentration of a particular nutrient in a given volume of soil reflects whether or not the nutritional level of that soil is adequate for optimum plant growth or production. Plant analysis is based in part on the concept that the amount of a specific nutrient in the plant tissue is related to the plant availability of that element in the soil. Plant analysis also reflects nutrient uptake conditions in the soil. Soil properties such as compaction, root disease, impervious layers and/or poor drainage may inhibit the uptake of nutrients by plants. Or a low concentration of one nutrient in the plant may result from the excessive application of another nutrient. Conversely, favourable soil physical properties and optimum soil moisture may accentuate nutrient uptake even though the soil may not have an abundant supply of nutrients.

As a result of these soil-plant interactions, there are certain instances when contradictions occur between soil and plant analysis results. For example, assume turf is growing on a soil in which the soil tests revealed a medium level of extractable Magnesium. A plant analysis from the area a few weeks later indicates that Magnesium is low. Immediately, the validity of the test results are questioned, which is an absolutely normal response. However, a closer examination of the plant analysis results revealed that the Calcium and Potassium concentrations of the turf were high.

Upon checking the information accompanying the plant analysis results, it was noted that Lime ($CaCO_3$) had been applied at 500g/m2 and a high rate of Potassium was applied in the fertilizer programme. As a result of these two management practices, the level of Calcium and Potassium in the soil was sufficient to reduce the uptake of Magnesium. This is one example of how soil testing and plant analysis can be used together for making better nutrient management decisions. Plant analysis can also be used to supplement a soil testing programme. It is particularly useful in distinguishing between Nitrogen and Sulphur deficiencies in turf, as deficiency symptoms of the two elements are similar. Plant analysis offers an excellent means of delineating which element is deficient (which cannot be ascertained through soil testing). If this distinction is not made properly and the wrong corrective treatment is applied, plant growth can be affected appreciably.

In the case of most turf grass species, a soil analysis prior to active growth in spring makes it possible to determine whether lime, dolomite, gypsum and/or specific fertilizer applications will be needed. Plant analysis of the turf during the growing season will indicate if the applied materials were effective and whether the predictions based on the soil analysis were correct.

AANDL EASTERN LABS-SAMPLING FOR PLANT ANALYSIS

Plant analysis assesses nutrient uptake while soil testing predicts nutrient availability. The two tests are complementary as crop management tools. Plant analysis will detect unseen hidden hunger and confirm visual deficiency symptoms. Toxic levels may also be detected. If it is done early, plant analysis will allow a corrective fertilizer application in the same season. A basic knowledge of plant structure is necessary before collecting samples. A leaf is made up of a leaf "blade" and a "petiole". The petiole is the stalk attached to the blade. A compound leaf may have several "leaflets" attached to it. In some cases, only terminal "leaflets" may be sampled, as in the case of walnuts and pistachios. A common error in tomatoes is when only leaflets are sampled instead of the whole compound leaf. This shows the importance of understanding proper sampling technique.

The most recent mature leaf (MRML) is the first fully expanded leaf below the growing point. It is neither dull from age nor shiny green from immaturity. For some crops, the most recent mature leaf is a compound leaf. The most recent mature leaf on soybean and strawberry, for example, is a trifoliate compound leaf: three leaflets comprising one leaf. For cotton, grape, potato and strawberry, petioles provide an additional indication of nitrogen status.

When sampling these crops, collect most recent mature leaves and their petioles. Detach leaves from petioles in the field to stop the translocation of nutrients. Put petioles in a separate bag. "Midribs" are the middle ribs to large leaves such as corn, lettuce, and cabbage, and would equate to a petiole sampling.

Deciding When to Sample

To monitor plant nutrient status most effectively, sample during the recommended growth stages for your specific crop Take samples weekly or biweekly during critical periods, depending on management intensity and crop value. However, to identify a specific plant growth problem, take samples whenever you suspect the problem.

The best time to collect samples is between mid-morning and mid-afternoon. Nitrate nitrogen varies with time of day and prevailing conditions but generally not enough to alter interpretation. Sampling during damp conditions is okay but requires extra care to prevent tissue from decomposing during shipping. Keep samples free of soil and other contaminants that can alter results. The appropriate part of the plant to sample varies with crop, stage of growth, and purpose of sampling. When sampling seedlings less than 4 inches tall, take whole plants from 1 inch above the soil line. For larger plants, the most recent mature leaf is the best indicator sample.

Taking A Representative Sample

Proper sampling is the key to reliable plant analysis results. A sample can represent the status of one plant or 20 acres of plants. In general, a common-sense approach works well. When problem solving, take samples from both "good" and "bad" areas. Comparison between the two groups of samples helps pinpoint the limiting element. Comparative sampling also helps factor out the influence of drought stress, disease, or injury. Take matching soil samples from the root zones of both "good" and "bad" plants for the most complete evaluation.

When monitoring the status of healthy plants, take samples from a uniform area. If the entire field is uniform, one sample can represent a number of acres. If there are variations in soil type, topography, or crop history, take multiple samples so that each unique area is represented by its own sample.

Choosing Sample Size

The actual laboratory analysis requires less than one gram of tissue. However, a good sample contains enough leaves to represent the area sampled. Therefore, the larger the area is, the larger the sample size needs to be.

Sample size also varies with crop. For crops with large leaves, like tobacco, a sample of three or four leaves is adequate. For crops with small leaves, like azalea, a sample of 25 to 30 leaves is more appropriate. For most crops, 8 to 15 leaves is adequate. For crops requiring petiole analysis, collect at least 15 to 20 leaves.

Submitting the Sample

Send the completed information sheet and proper fee with each sample. Use permanent ink or pencil on sample forms and bags Avoid numbering samples simply as 1,2,3 ... as it may lead to confusion later.

Give each sample a unique identifier that will help you remember the plants or area it corresponds to-such as HOUSE1, 15B, GOOD, or BAD. You can use up to six letters and/or numbers. Put the identifier on both the information sheet and the sample envelope. Pay attention to detail when filling out the information sheet. Note any conditions-drought, disease, injury, pesticide or foliar nutrient applications-that might be relevant. Indicate the analysis desired and provide very specific information on stage of growth and plant part if an interpretation is required. The laboratory does not automatically provide an interpretation, as some clients prefer to make their own.

Diagnostic interpretations require more details than predictive. When sending matching soil, solution, or waste samples, indicate the matching sample ID in the designated areas on the information sheet. Be sure the grower name and address are exactly the same on all matching information sheets. Ship all matching samples in the same container. Ship the tissue sample in a paper envelope or cardboard box so it can begin drying during transport. Samples put in plastic bags will rot, and decomposition may alter test results.

If samples are very wet, air-dry to a workable condition before packaging. Otherwise, decomposition or molding will occur. Include a completed plant analysis information sheet or cover letter with instructions within the same package. Processing will be delayed if sent separately. Also, include payment if you do not have an established account. Samples should be shipped by a carrier such as UPS or FEDEX, or by first class mail.

Interpreting the Report

Samples are analyzed the next day of their arrival. The prompt turnaround makes it possible for growers to take any corrective action needed to salvage the current crop. The report can be e-mailed or post mailed to the grower

PLANT ANALYSIS AS A GUIDE FOR MINERAL NUTRITION OF SUGAR BEETS

Nitrogen is the nutrient most likely to limit growth and sucrose production in sugar beets. Profitable responses to P fertilization are becoming common. Potassium fertilization is necessary in certain soils of the Sacramento-San Joaquin Delta. Sulfur is deficient in some soils of the upper Sacramento Valley. Micronutrient deficiencies of sugar beets are uncommon in California.

Research and practical experience have shown that plant analysis can serve as an effective guide in the fertilization of sugar beets. A plant analysis programme, properly carried out, will help optimize income for the producer, provide quality beets for processing, conserve fertilizer, and minimize the leaching of nutrients to underground water supplies.

CRITICAL CONCENTRATIONS

Critical concentrations of most of the essential nutrients have been determined for various parts of the sugar beet plant. These values along with concentration ranges found in healthy plants and levels in plants with deficiency symptoms.

The critical concentration for any nutrient is a relatively constant value which changds little with sugar beet variety, climate, or soil conditions. Nutrient concentrations, on the other hand, are changed by conditions that affect plant growth. For example, petioles of mature leaves of plants infected with the beet yellows virus may be higher in N03-N concentration than are healthy plants. This is because diseased plants grow more slowly, and nitrate accumulates in their petioles. Potential users of sugar beet leaf analysis sometimes express concern that differences in plant moisture content at different times of the day may significantly alter the concentrations of nutrients in plant tissues and thus affect the interpretation of analytical results.

In field experiments we observed little or no change in dry-weight N03-N concentration in petioles samples hourly from sunrise to sunset. Nor were there important differences with moderate (slight wilting) compared to no moisture stress. Only when plants were severely stressed was there a significant decrease in petiole nitrate. Plants moderately stressed compared to those severely stressed (loss of many leaves) had N03-N concentrations of 5, 1 00 and 3,600 ppm, respectively. Differences in the nutrient concentration due to variety, weather, moisture stress, disease and the like, are largely related to the kind and extent of root development or to other growth factors, and not to differences in internal requirements.

SAMPLING AND EVALUATION

For the sugar beet, recently matured leaves (the youngest leaves that have just attained maximum size) are usually the most appropriate part to sample. Petioles are most satisfactory for evaluating the plant's nutritional status for N, P, and Cl, while the blade is most satisfactory for the other essential elements.

A series of four samplings should be made throughout the growing season: one at thinning time, the second at early mid-season, the third at late rnid-season, and the fourth just before harvest. By comparing the analysis of these samples to critical nutrient concentrations, it is possible to detect and correct impending nutrient deficiencies before they occur, and to change the timing or method of fertilization to assure that fertilizer needs of the crop are fully met.

Large amounts of N are required to insure adequate top and root growth, but if storage roots are to be high in sucrose concentration the plants must be N deficient for 4 to 8 weeks before harvest. An estimate of the amount of fertilizer N needed for a specific crop should be made according to past history of the field or by soil tests. This amount should be applied to the crop early in the growing season, not later than thinning time. If an error is made, it is best to make it on the side of under-fertilization, because such a mistake can be corrected. If, however, the initital fertilizer application is more than the crop can use, fertilizer will be wasted, sucrose concentration (and, consequently, crop value) will be depressed, and excess nitrate will possibly be leached to the ground water.

Adequacy of the initial fertilizer application can be ascertained by analyzing petiole samples taken at 4 and 2 weeks before mid-season-the date midway between the dates of emergence and harvest. (To determine midseason for a crop to be overwintered, determine the date midway between emergence and November 1.) Indication of an impending deficiency of more than 10 weeks before harvest can be corrected by an addition of from 40 to 80 pounds of N per acre. These two early-season petiole samples can also contribute to the overall evaluation of the fertilizer programme as outlined above.

Abnormal Plant Growth

Frequently, plants in a field or part of a field show abnormal growth. A knowledge of nutrient deficiency symptoms gained from previous experience or a printed visual guide may lead to a tentative diagnosis which can be confirmed or discarded by analyses of comparable petioles and blades of abnormal and normal appearing plants. A useful visual guide is: Sugar Beet Nutrient Deficiency Symptoms, A Colour Atlas and Chemical Guide (Ulrich and Hills, 1969). To confirm a diagnosis of deficiency symptoms by chemical analysis, plants should be sampled soon after the symptoms appear. After a prolonged period of stress, the concentration of certain nutrients increases in physiologically mature tissue, especially if new supplies of the nutrient become available to the plant.

Guide to Harvesting

When there is a choice of fields to be harvested, the results of a systematic plant analysis programme can serve as a guide to scheduling of beet harvest. Those fields depleted of N first would be scheduled for an early harvest; those high in N would not be harvested until the N is depleted or would be held as long as possible before harvesting.

Root pulp as well as petioles may be analyzed to determine depletion of N before harvest. Pulp from preharvest root samples may be analyzed for nitrate by a semi-quantitative spot test using the diphenylamine reagent or, more accurately, by other quantitative methods. To maximize the sucrose concentration of beet roots, petioles and/or roots should be below critical N03-N levels about 4 to 8 weeks before harvest.

Nutrient Survey

A nutrient survey of sugar beet fields in an area can result in detection of deficiencies and identification of fields on which future fertilizer trials should be located. Surveys conducted by Farm Advisors or other interested agricultural groups can do much to improve fertilizer practices in an area.

Sampling Procedure

To serve as a measure of variability within a field, a minimum of two, but preferably four, samples should be taken from each field. From 25 to 50 leaves are usually sufficient to reasonably estimate the true mean of the nutrient concentration in the plants represented by the sample. A convenient sampling system is to divide a field into imaginary quarters. The sampler walks across the centre of each quarter at right angles to the plant rows and collects from 25 to 50 leaves per quarter of the field. Samples are placed in a paper bag of convenient size and taken to the laboratory for processing as soon as possible. Delays of more than 48 hours at ordinary temperatures should be avoided, and storage in closed compartments of automobiles during warm weather should not be allowed. If samples must be stored before drying they should be kept at 5°C; at this temperature, sugar beet petioles change very little in nitrate, phosphate, and K concentration. For microelement analyses, blades must be free of dust and should be washed for 30 seconds in a 0. I N HC1 bath followed by two successive rinses in distilled water.

A sample is reduced in size by cutting it into small pieces (approximately 1/8 inch), thoroughly mixing the pieces and then placing a sub-sample of about 100 grams (a large handful) of fresh material in a cut-down paper bag or other suitable container and drying it overnight in an oven, preferably with a forced draft, at 70'C to 80'C (158'- 1 76'F). After drying, samples are ground to pass a 40-mesh screen and stored. Samples treated in this way have been kept for many years without significant change in their mineral composition.

INTERPRETATION AND LIMITATIONS

With plants in the field, the likelihood of a growth response from the addition of a nutrient to the soil will depend upon whether the nutrient concentration of the plants is above or below the critical level. When the nutrient concentration is above the critical level and remains there throughout the growing season, there is little chance of a response in growth from the addition of more of the nutrient. When the nutrient concentration of the plants falls below the critical level, the chance of a growth response becomes greater the longer the plants remain deficient, provided other growth factors are favourable (*e.g.*, climate is favourable for rapid growth, and growth is not seriously limited by disease, water stress, and the like). If the deficiency appears late in the growing season, there is little chance for a significant yield increase with the addition of more of the nutrient as time is required for any deficiency to cause a measurable decline in crop yield. When plant material for a sample is taken from a large area, the sample may differ considerably in nutrient concentration within itself. When the average nutrient concentration of the sample is at or near the critical level, the material collected will represent both deficient and non-deficient plants. If there are areas in a field differing in plant-growth characteristics, these areas should be sampled separately.

Analysis of plant tissue usually reveals only one deficiency at a time. A second nutrient, or even a third, may be in short supply, but due to reduced growth caused by the primary nutrient deficiency, other nutrients will accumulate in the tissue. However, when the primary deficiency is corrected, the increased growth will decrease the concentration of the second nutrient, and most likely the second nutrient will become deficient rather soon. An exception to this general rule is with regard to P, deficient surgar beets which, particularly as seedlings, are also occasionally deficient in N, because P deficiency retards nitrate uptake. In the interpretation of the results for multiple sampling dates, the sharper the drop in nutrient concentrations and the earlier in the season it occurs, the greater the degree of deficiency. Even when the results of several sampling dates indicate the plants have been deficient for a considerable period, the exact degree of deficiency is not known. At harvest, however, the degree of deficiency for a given field can be estimated in terms of the maximum yield for the area, and fertilization for the next crop is then adjusted to this finding. For example, a field that produced only 25 per cent of the potential yield is much more deficient and will require more fertilizer than one that produced 90 per cent of its potential.

One of the greatest values of plant analysis is the prevention of deficiencies rather than their correction. Trends in nutrient content of leaves followed over a period of years can be a valuable guide in delineating the kinds, amounts, frequency, and methods of fertilization for more efficient crop production on a given field. Nutrient deficiencies can be predicted or at least detected as soon as they occur, and corrective measures can be taken before serious crop loss takes place.

4

Horticultural Plants

TROPICAL AND SUB-TROPICAL FRUITS

Tropical and subtropical fruits, in contrast with temperate fruits, can be broadly defined as those meeting all of the following criteria: crops that have their origin and commercial growing areas (when such exist) in the tropics or subtropics, plants that are evergreen and perennial, crops with a limited degree of frost resistance, and plants whose growth is practically non-existent below 50°F (10°C) (with some exceptions according to species and individual age).

A distinction between tropical and subtropical is possible if one considers that tropical species are not only sensitive to temperatures below 68°F (20°C) but indeed require a climate with average mean temperatures higher than 50°F (10°C) for the coldest month. Additionally most tropicals require humid environmental conditions.

Examples of truly tropical crops are traditional fruits native to Southeast Asia, like mangosteen, durian, and rambutan. A good example of a typical subtropical fruit crop is the cherimoya, which when cultivated in cold subtropical areas may suffer some foliage loss during the winter with regrowth in spring. However, some fruit crops can be cultivated equally well in either the tropics or the subtropics, of which the banana and the avocado are the most outstanding examples.

BANANA

Banana (Musa paradisiaca L.) occupies over 1,64,000 hectares, mainly in Tamil Nadu, West Bengal, Kerala, Maharashtra, Gujarat, Karnataka, Assam, Andhra Pradesh and Bihar.

Though some inferior types of banana are found growing as far north as the Himalayas, its commercial importance is mainly limited to the more tropical conditions, such as those prevailing in central, southern and north-eastern India. It is a moisture- and heat-loving plant and cannot tolerate frost or arid conditions.

Varieties

Cultivated varieties are broadly divided into two groups: table and culinary. Among the former are 'Poovan' in Madras (also known as 'Karpura Chakkarekeli' in Andhra Pradesh); 'Mortaman', 'Champa' and 'Amrit Sagar' in West Bengal; 'Basrai', Safed Velchi', Lal Velchi' and 'Rajeli' in Maharashtra; 'Champa' and 'Mortaman' in Assam and Orissa; and 'Rastali', 'Sirumalai', 'Chakkarekeli', 'Ney Poovan', 'Kadali' and 'Pacha Nadan' in southern India. 'Basrai', which is known under different names, *viz*. 'Mauritius', 'Vamankeli', 'Cavendish', 'Governor', 'Harichal', is also grown in central and southern India. Recently, the 'Robusta' variety is gaining popularity in Tamil Nadu and Karnataka.

The 'Virupakshi' variety (Hill banana) is the most predominant variety in the Palni Hills of Tamil Nadu. Among the culinary varieties, Nendran bananas, 'Monthan', 'Myndoli' and 'Pacha Montha Bathis' are the leading commercial varieties in southern India. 'Gros Michel' is a recent introduction into southern India; it is suitable for cultivation only under garden-land conditions and is generally fastidious in its cultural requirements. It is not, therefore, in favour with the cultivation.

Propagation and Planting

Propogation is by suckers or off-shoots which spring at the base of a banana-tree from underground rhizomes. Vigourous suckers, with stout base, tapering towards the top and possessing narrow leaves, are selected for plant. Each sucker should have a piece of underground stem with a few roots attached to it. Banana suckers can be planted throughout the year in southern India, except during summer, whereas in the rest of the country, the rainy season is preferred. They are planted in small pits, each just enough to accommodate the base of a sucker. The planting-distance varies from 2m×2m in the case of dwarf varieties to 4m×4m in the case of very tall varieties.

Manuring

An application of 20 to 25 kg of farmyard manure, together with about 5 kg of wood-ashes per plant is given at planting time. In southern India, ammonium sulphate is applied one month, five months and nine months after planting 20 kg per hectare each time.

In western India, a little over 2 kg of oilcake per stool is applied during the first three months after planting. A complete fertilizer mixture may be applied to supply 100 to 200 kg of N, 100 to 200 kg of P_2O_5 and 200 to 400 kg of K_2O per hectare. A green-manure crop is also considered beneficial. Trials at the Indian Institute of Horticultural Research have shown that for the 'Robusta' variety, a fertilizer mixture comprising 180 g of N + 108 g of P_2O_5 + 225 g of K_2O per plant is ideal.

After-Care

The removal of suckers, dry leaves and pseudostems, from which the fruits have been harvested, constitute the main after-care. Daughter-suckers should be removed promptly until the mother-plant flowers, when one daughter-sucker may be allowed to take its place.

The removal of dry leaves and useless pseudostems requires to be done in time. After all the fruits are formed, the pendant portion of the remaining inflorescence along with the heart should be removed. The propping of plants with bamboo poles, especially those which have thrown out bunches, is necessary wherever damage by wind is apprehended. Where the wind damage is recurring, dwarf varieties should be preferred.

Irrigation

The banana-plants require very heavy irrigation. Irrigation is given in most places once in seven to ten days. Stagnation of water in the soils is not very congenial to the proper growth of banana and, hence, the drainage of soil is also essential.

Harvesting

Early varieties commence flowering in southern and western India about seven months after planting, and the fruits take about three months more to ripen. In the Andhra Pradesh delta areas, the fruits are ready for harvesting about seven to eight months after planting.

The first crop of the 'Poovan' variety matures in 12 to 14 months and the second in 21 to 24 months after planting. In other parts of India, the first crop is usually gathered a year after planting, whereas the succeeding crop may be ready in six to ten months thereafter. The bunch is harvested just before it attains the ripening stage.

When the fruits have reached the full size, they become plump, and mature with a distinct change in colour. For long transport, the bunch may be harvested somewhat earlier. The bunch is cut, retaining about 15 cm of the stem above the first hand. The yield varies considerably from 26,000 to 55,000 kg per hectare.

Curing and Marketing

The ripening of banana is done in several ways, *e.g.* exposing the bunches to the sun, placing them over a hearth, wrapping them in closed godowns or smoking them in various ways. One of the common ways is to heap the fruits in a room and cover them with leaves, after which fire is lit in a corner and the room is closed and made as air-tight as possible. Ripening takes place usually in 30 to 48 hours. In a cool store, the bunches ripen well at about 15° to 20°C. The application of vaseline, a layer of clay or coal-tar to the cut-ends of the stalks prevents rotting during ripening and storage.

Wrapping up the fruits and packing them in crates help to reduce the damage during transport.

MANGO

Mango (Mangifera indica L.) occupies nearly half of the total area under fruits in the country. It is adaptable to a wide range of soil and climatic conditions and grows well right from Assam to the southern-most limits of the country and from the sea-level up to about 1,500 metres. It withstands both fairly dry conditions and heavy rainfall, provided severe and recurring frosts in winter do not endanger the young trees.

Varieties

The number of varieties is very large. Each variety has its own peculiar taste, flavour and consistently of pulp. Some of the important commercial varieties grown in different regions are: 'Bombay yellow', 'Alphonso', 'Gopal Bhog', 'Zafran' (all early), 'Langra', 'Desheri', 'Safeda Lucknow', 'Safeda Malihabad', 'Fajrizafrani' (all mid-late). 'Fajri', 'Same Bihisht', 'Chausa', 'Taimura' (all late).

In Uttar Pradesh; 'Bombai', 'Alphonso', 'Hemsagar', 'Krishna Bhog', 'Aman Dasheri', 'Gulab Khas' (all early), 'Langra', 'Aman Abbasi', 'Khasul-Khas' (all mid-late), ' Sinduri', 'Sukal', 'Taimuria' (all late) in Bihar; 'Bombai' or 'Maldah', 'Gopal Bhog', 'Hemsagar' (all early), 'Krishna Bhog', 'Zardalu' (both mid-late), 'Murshidabadi', 'Fazli Maldah' (both late) in West Bengal; 'Alphonso', 'Pairi', 'Cowsji Patel', 'Jamadar' in Bombay; 'Swarnarekha', 'Benishan', 'Cherukurasan', 'Panchadarkalasa', 'Desavathiyamamidi', 'Sannakulu', 'Nagulapalli', 'Irsala' in Circars; 'Rumani', 'Neelum Benishan', 'Bangalore', 'Alampur Benishan' in Rayalaseema; 'Murshidabadi', 'Mulgoa', 'Goabunder', 'Benishan', 'Neelam', 'Totapuri' or 'Bangalora' in Telengana; 'Alphonso', 'Peter', 'Rumani' in central districts; 'Mundappa', 'Neelam', 'Alphonso', 'Olour', 'Benn*et Al*phonso', 'Kalepad', 'Peter', 'Fernandin' in Coorg and Karnataka; and 'Padiri', 'Alphonso', 'Peter', 'Neelum', 'Bangalore', 'Rumani' in Tamil Nadu. In Goa, some excellent varieties like 'Alphonso', 'Fernandin', 'Mankurad' and 'Moussorate' are under cultivation.

The new mango variety, 'Mallika' evolved at the Indian Agricultural Research Institute is now gaining popularity. Other varieties, such as 'Jehangir' and 'Himayuddin', produce high-quality fruits, but are poor in yield and cropping tendencies. Attempts are being made to evolve hybrid progenies by crossing.

Propogation and Planting

Propogated vegetatively by inarching or budding in situ in the nursery, either by using Forkert or by using the T-method. The beginning of the monsoon in light-rainfall areas and the end of the monsoon in heavy-rainfall regions are the most suitable periods for inarching or budding. Recently, veneer-grafting has been found to be the best method of mango propagation. Grafted plant are ready for transplanting in the field after six to twelve months. Select straight-growing grafts and set them in pits filled with soil mixed with farmyard manure (45 kg) and a fertilizer mixture containing 0.225 kg of N, 0.45 kg of P and 0.225 kg of K per pit. The planting-distance is 7.5 to 9 metres in poor shallow soils and 15 to 17 metres in deep fertile soils. The beginning of the monsoon in low rainfall areas or the end of the monsoon in heavy rainfall tracts is the best time for planting. The graft-joint should be at least 15 cm above the ground.

Pruning

No systematic pruning is done. The removal of dead-wood and the thinning of over-crowded and mis-shapen branches after about four years is all that is necessary; flowers that appear during the first three or four years should be removed.

Culture

Before planting, the field is ploughed, harrowed and levelled. Thereafter, it is ploughed and harrowed twice a year, once in the beginning of the monsoon and again at the close of the rainy season or in the cold-weather. It is green-manured once every two or three years. Short-season intercrops, like vegetables, may be taken during the first four to five years. Young plants require irrigation regularly. After five to six years, when they have established themselves, the trees are able to grow and fruit satisfactorily without irrigation in most parts of Peninsular India.

In northern India, they have to be irrigated throughout their life. Irrigation is usually withheld during the cold weather before flowering, especially in deep retentive soils. Though the exact manurial requirement is not known, regular manuring is beneficial. The dose recommended for the bearing trees is 45 to 70 kg of farmyard manure, 0.5 to 0.7 kg of N, 0.7 kg to 1.0 kg of P and 1.2 to 1.5 kg of K per tree. Nitrogen and half of potash may be given before the monsoon, and farmyard manure, phosphate and half of potash in October or before flowering starts.

Crop Irregularity

Grafted mango-trees bear fruits from the fourth or fifth year onwards and a full crop from the tenth or fifteenth year. The erratic bearing of mango is well known. It depends upon the variety, the weather and climatic conditions and cultural treatments. The selection of regular-bearing varieties, timely cultural practices and proper nutrition help to produce a regular crop. New growth in spring, on which flower-buds are produced during the next winter, can be encouraged by applying nitrogenous fertilizers (0.45 to 0.90 kg of N per tree). In the case of heavy late rains, an additional ploughing in winter helps to produce flower-buds in January-February. In the case of individual trees, ringing or girdling in August-September may also to help to force flower-buds the following winter. The application of Ethral (200 ppm) from September onwards has been found to induce flowering in mango in Karnataka by the Indian Institute of Horticultural Research.

Improvement of Old and Seedling-Trees

Mango-trees of inferior varieties, so also those raised from seeddlings, can be converted into choice varieties by grafting them in situ either by crown or side-grafting. In crown-grafting, the trunk of the tree is cut down to about half a metre from the ground and one or more scions of the selected variety are inserted into it between the bark and the wood by splitting open the bark. The scion should be a dormant, terminal shoot, about 12.5 mm in diameter, with a whorl of plump swollen buds at the top. In side-grafting, the procedure is the same as in crown-grafting, except that the trunk of the stock tree above the grafting joint is cut down after the scions have sprouted and have established themselves properly.

Old trees, having several branches, can be similarly improved (top-worked) by crown-grafting on each branch at a suitable height. Sometimes, the grafting is done by inarching, but the process is cumbersome, expensive and not very satisfactory.

Harvesting and Marketing

The fruit takes five to six months to mature. Depending upon the onset of flowering, the mature fruits are ready for harvesting from April to May in western India, from May to June in the Deccan, from February to March in Malabar, from April to July in the coastal Andhra Pradesh, from May to August in Mysore and Rayalaseema, and from June to August in northern India. The mature fruits are harvested by severing the stalks to which they are attached, when they are still green and hard. The signs of maturity vary with different varieties. As a mango tree usually bears flowers in three or four distinct flushes lasting over a month, it is preferable to harvest the fruits as they mature. The fruits, so harvested, can be transported after packing them in baskets or wooden crates, properly padded with straw, wood-shavings or wool, to long distances. For overseas markets, they are packed in a single layer in specially designed wooden crates.

For ripening, the fruits are spread out on rice straw in a single layer. Two or three such layers are built one above another in a well-ventilated room. The mangoes are ready for disposal after they change colour. Yield varies considerably with the variety, vigour of growth, flowering, etc. A grafted tree yields about 300 to 500 fruits in the tenth year, about 1,000 in the 15th year and 2,000 to 5,000 from the 20th year onwards.

ENVIRONMENT AND HORTICULTURAL PLANTS

Horticultural plants grow in response to their environment. Plants are cultivated from many areas of the world. Each has a different set of environmental conditions to which the plant has become adapted. Much of horticulture is concerned with modifying plant environments to enhance growth. This chapter covers several of the major environmental factors that influence plant growth and the influence of climatic variations on horticultural plants.

CLIMATE

Climate is the combined effects of temperature, light and elements of moisture such as clouds, rain, hail, snow and wind. Climate usually refers to long term weather patterns in a region or as it is often called, macroclimate. Temperatures are greatest at the earth's surface near the equator where the sun's rays strike directly.

As you move north or south from the equator it gradually becomes cooler. Altitude also influences temperatures with cooler temperatures occurring at higher altitudes. Climatic patterns are also altered by large bodies of water nearby. More energy is required to raise the temperature of water than air. Water releases this energy as it cools. Thus, water acts as a buffer for nearby land areas and reduces temperature extremes. Inland areas of the U.S. lack the protection of large bodies of water and experience high summer temperatures and cold, arctic fronts in the winter.

Mountains also alter climate by blocking natural wind and moisture movement. Several areas of the Pacific northwest have wet, rainy areas on the ocean side of mountains and dry areas on the inland side. Vegetation also influences climate. Forested areas have a higher relative humidity.

This and additional evaporation of water from tree leaves has a cooling effect. Human activities also influence climate. Large reservoirs have been constructed in many areas. Metropolitan areas are warmed by combustion used in heating and by buildings, streets and other structures absorbing and holding heat. Dust, environmental pollutants and smoke also alter the climate. A concern in many areas of the eastern U.S., Canada and Europe is acid rain, formed as rainfall absorbs acidic materials from air pollution. All of these macroclimate differences have an effect on plant species growing in an area and may cause climatic variations.

Microclimate refers to the climate in a small area such as near a plant or in a yard or garden. Cold air is heavier than warm air and tends to settle in low areas creating a temperature inversion. Temperatures in low areas can be several degrees cooler than surrounding hillsides. These pockets of cold air are called frost pockets. Orchards are generally located on hillsides to avoid areas likely to damage flowers in early spring. Selecting a home site on a sloping area can also avoid low temperature injury to ornamentals in the fall and spring.

Carefully locating plants around a home can take advantage of microclimate differences. Shade loving plants can be grown on north or east exposures and plants can be protected from the blast of west or north winds by planting them on the protected side of the house. Horticulturists need to be aware of the influence of climate in a particular area, both macroclimates and microclimates. Selection and growth of horticultural plants is directly influenced by climate.

TEMPERATURE

Temperature largely determines what can be grown in a particular area. Photosynthesis (the conversion of solar energy for use by the plant) and respiration (the using of stored energy compounds in the plant for growth and development) are regulated by temperature. With most biological reactions each 10°C (18°F) increase in temperature doubles the rate of the reaction.

Therefore, we expect the rate of growth at 68°F to be twice as fast as at 50°F. This relationship is known as Q10. Each plant has an optimum temperature for growth and development . There is a maximum temperature where plant growth stops and permanent injury to the plant occurs. Likewise, there is a minimum temperature where plant growth stops and freezing or chilling injury occurs.

Cool Season Plants

Depending on their area of native adaptation, plants prefer cooler or warmer temperatures. Cool loving plants tolerate temperatures slightly below the 32°F. Growth is poor at warmer temperatures. Examples of cool season plants include vegetables (cabbage, broccoli, lettuce, radishes, peas), flowers (crocus, daffodil, tulips, violets), fruits (apples, pears, plums, American grapes) and shrubs (forsythia, lilac, spirea, honeysuckle). Cool season plants prefer average daily temperatures below 70°F. Premature seed stalk formation or bolting of biennial vegetables is a problem for cool season crops. Cool temperatures following planting compresses their natural growth cycle and seed stalks often form later in the season.

Warm Season Plants

Some plants prefer temperatures above 70°F and are usually injured by freezing temperatures. Sometimes these plants are called non-hardy or tender plants, while those that can withstand freezing temperatures are called hardy plants. Examples of warm season crops are vegetables (beans, tomatoes, sweet corn, melons), flowers (roses, lilies tropical foliage plants) and fruits (peaches, apricots, sweet cherries).

Freezing Injury

When the temperature drops below 32°F there is a chance of injury to some plants. This is frequently called frost injury. Frost is ice formation on a surface while plant injury occurs only when ice forms within the plant. Generally, the period just before dawn is the time when the temperature reaches the lowest point and freezing injury occurs.

Freezing injury results in an immediate water soaked, blackened appearance that quickly turns brown and dies. It is a common horticultural practice to cover plants with insulating materials or sprinkle plants with water to prevent freezing temperatures from occurring within the plant. While water from a sprinkler may feel very cold it has some heat energy that is passed on to the plant.

There is a lot of current research on treatments to allow plants to be exposed to lower temperatures without injury and to understand the mechanism of freezing injury to plant tissues.

Chilling Injury

Injury can occur to warm season plants or products exposed to low temperatures. This low temperature injury is called chilling injury and results from a malfunction in the normal plant growth processes rather than the freezing of water within the plant.

Crops such as cucumbers, tomatoes, tropical fruits and most tropical foliage plants, are subject to chilling injury. The development of chilling injury usually results in rapid respiration, the development of rots and molds, bitter or "off flavours" and abnormal colours. Chilling injury may occur at temperatures below 45-50°F in many plants or products. You see it when a banana turns black after being placed in a refrigerator.

Rest Period

There are periods when a plant or seed does not grow despite favourable conditions. These periods are called rest periods. They usually coincide with the natural period of dormancy for perennial plants during the winter. During this time deciduous plants lose their leaves and herbaceous plants are frozen back to the ground.

Buds of woody plants do not grow due to physiological and biochemical processes. Following a period of cold, the buds break (end) their rest period and continue growth. The temperature and hours needed to break dormancy has been determined for many plants. Fruit varieties are grouped by their requirement of hours at or below a critical temperature (usually 40°F) to break the rest period.

Desirable environmental conditions are necessary to resume growth following the breaking of the rest period. Many temperate plants produce seeds with a natural rest period following ripening. This rest period can only be broken by exposure to cold temperatures for a period of time. This process is called stratification. We can artificially stratify seeds by storing them in a moist medium such as peat moss or sand at near freezing temperatures (usually 40°F) for a number of weeks. This breaks the normal rest period and allows the seeds to germinate uniformly.

Woody Plant Survival

An example of hardiness has already been provided for annual plants tolerant to low temperatures. Woody plants can also tolerate low temperatures to varying degrees.

Native plants have, through the years, adapted to withstand low temperatures and are said to be hardy to the area where they evolved. The level of low winter temperatures that a particular type of plant can withstand has been determined for most woody plant species. The United States is divided into areas called plant hardiness zones.

The zones are numbered 1-10 from colder to warmer areas. This map was developed by the US Department of Agriculture. Woody ornamental plants survive average normal low winter temperatures in these areas and plants can be characterized by their adaptation to these climatic zones. It must be remembered that these zones are based on the average of all of the lowest winter temperatures recorded over a 50 year period. In certain extreme years, such as in January 1994, severe damage to plant material occurred because the temperatures were so much colder than normally would be expected. Damage also frequently occurs to normally hardy plants when an unusually early frost or freeze hits plants before they are fully dormant or in the spring after they have begun to lose dormancy.

Modifying Temperatures

Much of horticulture is involved in altering the environment to improve plant growth. Areas of horticultural production have developed in areas where natural protection from temperature extremes exists. Such protection is provided by large bodies of water such as the western area of Michigan and the western areas of New York bordering the Great Lakes.

Buildings, streets, etc. in large cities absorb heat and reradiate it at night providing several degrees of warmth against the first freezing temperatures of the fall.

Growing plants in a managed environment like a greenhouse was developed to provide a favourable temperature during adverse periods. There are other examples of treatments horticulturists use to modify temperature. These including the use of mulches.

Black or clear plastic absorbs heat and transfers it to the soil encouraging earlier plant growth in the spring. Hotcaps or hot-tents - paper or plastic covers also serve as miniature greenhouses to modify the microclimate around transplants in the spring. Sheets of fibreglass or porous plastic are frequently placed over rows of plants to warm them in the spring. This type of structure is called a cloche (pronounced klosh) or row cover. Another example or a specialized plant growing structure is a hotbed or coldframe used to start seedling plants. Water mist is frequently used as a means of reducing freezing temperature injury.

The protection provided by water mist is based on the principle that as water freezes it gives off heat. Thus, as the water freezes it gives off enough heat to protect plants for a few degrees. Other examples of frost protection include smudge pots, stoves and air turbulators used in orchards to force warmer air around plant surfaces. Horticulturists continue to study ways to understand the mechanisms of temperature injury, ways of improving the adaptation of plants to temperature extremes and ways of altering the temperature around horticultural plants to improve their growth.

WATER

A brief drive through the desert, failure to water a pot of flowers, or wilted lettuce on a salad bar reminds us of the importance of water for plant growth and development. Just as with temperature extremes, plants have adapted to grow in areas of low water availability. Thick, fleshy leaves, layers of wax, specialized structures for water storage and other mechanisms help plants survive in areas of limited water. Such plants are called xerophitic.

Water in the Plant and Soil

Water constitutes an important part of plants. Living plants need a constant flow of water from the roots to leaves to continue their life processes. Water in plant cells provides pressure on cell walls, called turgor pressure, keeping the plant rigid. Water is an important constituent of photosynthesis. Water flows from the root, up the stem and into the leaves in specialized conducting cells called xylem cells. Movement of nutrients and other constituents flow in water in the plant as well.

Pressure from water in the cell enables cells to enlarge and expand. Water not used by plant cells evaporates and moves out of the leaf through small pores or openings in the leaf surface called stomata. Evaporation of water from the leaf is called transpiration. Evaporation of water from the leaf has a cooling effect and reduces temperatures in the surrounding environment. Water moves into root cells from the soil.

Soils differ in the amount of water that they can hold. Sandy soils hold less water than loam or clay soils. Water available to a plant from a particular soil is called the soil water potential. Some water is held so tightly by the soil that the plant cannot extract it. Water that the plant can extract from the soil is called available water. The root system develops in the soil to obtain water for the plant. Sudden flooding of the soil cuts off oxygen to the root system and causes root death. Plants in flooded locations can die quickly from such injury.

Too much water can be as serious a problem as not enough water. For this reason, a greenhouse manager must carefully monitor the water level in containers to provide the proper amount of water to meet the crop's needs. In field situations, drainage tiles or pipes are often installed to allow excess water to be removed. Water is lost from soils by evaporation from the soil surface. Evaporation of water from soil combined with loss of water through the plant (transpiration) is referred to by the combined word evapotranspiration, or loss of water by plant and soil combined.

Applying Water to Plants

Horticultural production has developed where water is abundant. Applying water to supplement rainfall is a standard practice for many horticultural crops. Irrigation should replace water removed by plants. Considerations in irrigation amounts and frequency depend on the needs of the crop, the depth of roots and the ability of the soil to absorb water.

A certain portion of the water is lost to evaporation before it reaches the soil. On a windy day a sprinkler system putting out a fine mist may lose 25-30 per cent of the water applied before it reaches the soil. Drip or trickle irrigation allows small amounts of water to be applied frequently to a portion of the root zone. This reduces the amount of water needed to grow a crop.

Effects of Limited Water Availability

For many horticultural plants there is a period of critical water need when water may not be available. For most plants this is during the period of flowering and fruit development. A symptom of water need is wilting of the foliage. Plants are affected by stress before the external symptom of wilting occurs.

Emphasis on water levels necessary to support plant growth and scheduling irrigation to minimize water use is important. Water needs can be predicted from climatic factors such as temperature, humidity, wind and solar radiation combined with particular characteristics of the crop and soil type. Microcomputers assemble various weather factors and develop predictions and schedules for adding supplemental water. Fluctuations in water available can cause sudden splitting of cherries, tomatoes and cabbage. Uniform water supplies are necessary to prevent these problems.

LIGHT

Light is essential for photosynthesis. Green plants are the original solar collectors. They transfer energy from the sun to a usable form of energy for the plant. Light varies in intensity, quality and duration or photoperiod.

Intensity

The intensity or amount of light available has a direct influence on the amount of photo-synthesis that can occur. Some plants have adapted to growing in areas of low light availability such as a jungle floor. This is why tropical foliage plants grow well in limited indoor light. Grasses are native to the open plains and require high light intensity for growth. When light intensity is lowered, the rate of photosynthesis is reduced. When the level of photosynthesis equals respiration, it is referred to as the compensation point. Below compensation point growth ceases.

Sudden exposure to high light intensity can cause leaf scorch or sunburning. The symptoms are large, brown, dead areas on leaves. Plants growing in low light have a thinner layer of wax on leaf surfaces. This helps the leaf to capture more light. However, moving it suddenly into bright light results in rapid water loss. High light intensity can also cause fading of flowers, especially in hot summer weather. An area of special interest to the greenhouse industry is the use of high intensity lights to supplement natural light and to extend the photoperiod. Crops of high value make the expense of installing and operating high intensity supplemental lights economical.

Light Quality

Light quality refers to the "colour" of the light or the portion of the light spectrum. The biochemical constituents of chlorophyll, the compound responsible for photosynthesis, absorb only light from particular portions of the light spectrum.

Artificial lights supply light of particular portions of the spectrum. Light absorbed by plants is from the blue-violet and orange-red parts of the spectrum. Specialized lamps have been developed to provide light rich in wavelengths from this part of the spectrum. These lamps have the additional benefit of creating dark green foliage and a deep, intense flower colour. Excellent plant growth can be achieved with ordinary fluorescent bulbs. Other portions of the spectrum such as ultraviolet light are important in the coloration of some fruits and the development of autumn colour in leaves.

Photoperiod

The daily duration of light is important for many plants. The lengths of light and dark period in a 24 hour cycle influences the blooming response of many plants. Some plants form flowers with shorter day lengths (usually less than 12 hours).

These plants are called short-day plants and grow without blooming during days with long photoperiods. Other plants are long-day plants and form flowers during longer days (usually 14 hours or longer). The most common examples of short-day plants are poinsettia and chrysanthemum that flower during the shorter days of autumn. Examples of long-day plants include many of our summer flowering annuals. Greenhouse producers must know the photoperiod requirements of plants and alter the day length to induce flowering at the proper time. Days are shortened by covering greenhouse benches with dark shade cloth. Day length is extended by artificial lights over the benches. Many plants are day neutral. These plants are not influenced by the number of hours of light. A common example is the tomato that blooms and fruits with favourable temperatures during any day length.

While we speak of day length, it is important to remember that it is really the length of the dark period that makes the difference. To keep a shortday plant from flowering naturally in the fall we only have to break up the dark period into two shorter segments. For some plants as little as five minutes of light in the middle of the night will keep it from flowering.

Relative Humidity and Wind

Relative humidity is the amount of water vapor in the air. It is closely related to temperature. As the temperature goes down, relative humidity goes up. Relative humidity is measured as a percent. When it reaches 100 per cent we have rain, snow, fog, or sleet. This is why it rains as a cold front goes through.

As the relative humidity drops, evaporation increases. Plants leaves give off more water at 40 per cent than 80 per cent relative humidity. This water loss increases if the wind is blowing. Tropical foliage plants are native of humid, jungle environments and often suffer in the dry, indoor air in winter. Grouping plants together and placing plants on a layer of pebbles in standing water helps increase relative humidity. Lowering the interior temperature of homes is beneficial in reducing some of the low humidity problems with plants.

Humidity is often related to the development of certain plant diseases. Warm, damp conditions are ideal to encourage the growth of fungi. Many fungal spores must have water to begin growing. Damping off is a common problem for seedlings. It results from fungi rotting small plants at the soil line. Increasing air circulation to lower humidity and reducing watering can generally stop the problem.

Wind is a problem with plant production in some areas. In addition to blowing away valuable soil, wind can also cause injury to tender plant stems by sandblasting the stems at the soil line. New transplants and newly planted trees are often shaded as a protection from hot sun and wind. In many areas of the Great Plains of the U.S., the planting of rows of trees in a windbreak protects homes and animals. Strong winds in storms damage many trees and shrubs each year. Wind damage is especially severe when it is in combination with ice. The plant species is also important. Trees such as silver maple, Siberian elm and willow are more frequently injured by wind than oak, walnut and sugar maple.

Interrelationships

Plants are influenced simultaneously by all segments of their environment. As light intensity increases, temperature generally increases with a corresponding increase in photosynthesis and water loss by the plant. Many times horticulturists must alter one environmental factor as another changes. When light levels are reduced, a greenhouse manager will often lower temperatures. Without the temperature reduction, plants become tall and spindly. The reduction in temperature limits the problem. Although a lot is known about environment factors that influence horticultural plants, much is still to be learned. Urban conditions influence plants in both positive and negative ways. Learning to compensate for these negative influences will become more important in the future.

CONCLUSION

Horticulture is a broad field of study encompassing the production, maintenance, sales, and service of bedding, flowering, ornamental, and nursery plants and food crops. This major has four distinct concentrations. Floriculture Crop Production stresses greenhouse production. General Horticulture allows students to design their own focus of study.

Retail Floral Design emphasizes the principles and practices which prepare students for the successful operation of a retail floral business by concentrating on design, handling of plants, and business practices. Management of Woody Plants covers plant production and management in a nursery and the operation of a retail garden centre.

Increasingly precise management of horticultural crops lends itself to the use of mechanistic crop simulation models to predict growth and yield. In this chapter, we encourage systematic approaches for developing, parameterizing, and testing of dynamic crop growth models for horticultural crops. Using the *Cropgro* model as an example, we describe crop development, photosynthesis, leaf area expansion, as well as the addition, growth, and maturation of individual fruits.

An algorithm is described for predicting fruit fresh weight, size, and dry matter concentration from dry matter growth and physiological age of individual tomato (Lycopersicon esculentum Mill.) fruits. Parameterizing crop models for response to climatic factors, especially temperature, requires interpretive use of literature information, along with testing against growth data from a range of growth conditions. A review of tomato literature for temperature dependencies of processes is presented with a view to how this information would be used to parameterize the CROPGRO-tomato model.

In modern horticulture, growers are searching for reliable biosensors and diagnostic tools to evaluate plant eco-physiological behaviour in order to optimize plant growth and crop development. One of these biosensors is the linear variable displacement transducer (LVDT) which accurately measures stem diameter variations. The variations of stem diameter reflect the combination of two main phenomena: irreversible radial stem growth and daily shrinkage and swelling of the elastic stem cells due to changes in internally stored water. A correct interpretation of the composite LVDT-signal, with respect to its practical application, requires an unambiguous distinction between these two components of the signal. This chapter describes a comprehensive model for stem diameter variations which includes a mechanistic description of radial stem growth and elastic stem diameter changes related to changes in internally stored water.

Using the results of a two-year-old beech tree (Fagus sylvatica), the practical application of the model is illustrated with measured LVDT-signals where radial stem growth and elastic changes of the stem could be distinguished. As this approach is also applicable to other ornamental or vegetable crops, the model can be considered as a very powerful tool for interpretation of the composite and complex origin of the variations in stem diameter as is observed with LVDTs.

5

Trends in Horticulture Growth

PROGRAMME IMPLEMENTATION

- Thrust on Hi-Tech Horticulture and precision farming.
- Expansion of area under micro-irrigation and fertigation.
- Stabilizing the area of water loving crops and expanding the area under dry land crops with a focus on effective water management.
- Area expansion of horticulture crops through various Waste Land Development programmes.
- Strengthening the system of production of pedigree and hybrid planting materials in both public and private sectors.
- Promotion of Organic farming with focus on export market.
- Promotion of Agri Export Zones (AEZ) for specified crops.
- Building up of Public and Private Partnership.
- Promotion of Contract/ Corporate farming.
- Empowerment of farmers with special focus on farm women.
- Effective transfer of technologies by tour-cum-training to farmers.
- E-Governance and Human Resources Development through effective training for extension officers.
- Linkage with Agro Processing Industries with New Anna Marumalarchi Thittam (NAMT).
- Post harvest management and reduction of post harvest losses.

TAMIL NADU HORTICULTURE DEVELOPMENT MISSION

The Government have setup a Mission for Horticulture development in Tamil Nadu vide G.O. Ms. No.155, Agriculture (H1-2) Department, dated 28.4.2003 with a mandate to give impetus to production, processing for value addition and marketing of Vegetables, Fruits, Flowers and Medicinal Plants. Tamil Nadu is the first state in India to setup a separate Mission for development of Horticulture in the year 2003. It is aimed to double the production of horticulture crops by the year 2011-12 in Tamil Nadu. The present level of productin of 99.47 L.MT will be doubled to 187.80 L.MTs during the year 2011-12.

OBJECTIVES OF THE MISSION

- Improving production through balanced nutrition management.
- Evolving suitable mechanism for regulating the production of quality planting materials and giving impetus to need based research.
- Establishing adequate infrastructure for post harvest management especially preservation and marketing.
- Encouraging active involvement of farmers associations in adoption of modern technological practices.

A Governing Body has been constituted to monitor the effective implementation of the mission under the Chairmanship of the Honourable Minister for Agriculture with the Vice- Chancellor of Tamil Nadu Agricultural University, Secretaries to Government and Senior officers of various departments and Nominees of Central Government agencies like National Horticulture Board, APEDA (Agricultural and Processed Food Products Export Development Authority) as members. The Chief Secretary to Government is the Vice Chairperson of the Governing body.

Mango Development Mission

Under Mango Development Mission it has been programmed to layout demonstration plots in the farmer's field to enable them to have a visual impact of various new technologies involved in the cultivation of Mango crop. A sum of ₹.22.55 lakhs has been proposed for this purpose. National Horticulture Board has agreed to sanction ₹ 10.00 lakhs out of which released a sum of ₹.5.00 lakh for implementation of this scheme. On receipt of Government Order, this scheme will be implemented in the selected Mango growing districts.

Cashew Development Mission

Under Cashew Development Mission it has been programmed to layout demonstration plots in the farmer field to increase the productivity of cashew with high yielding varieties and hi technologies. A sum of ₹.23.00 lakhs has been proposed for this scheme. National Horticulture Board has agreed to sanction ₹.10.00 lakhs out of which released a sum of ₹ 5.00 lakh for implementation of this scheme. On receipt of Government Order this scheme will be implemented in the selected cashew growing districts.

Training to Farmers

With a view to expose the farmers to various hitech aspects of Horticulture, farmers are taken to various research stations in out of state on tour-cum-training basis. About 10,000 Nos. of farmers are given training during the year 2004-2005. Apart from this farmers were given awareness in various horticulture technologies of Medicinal Plants, Flower cultivation and Vegetable Cultivation through the conduct of District level seminars and Exhibition.

The extension officers and field level workers of this department are also given training in the latest technologies of horticulture from the Research Stations situated inside and outside the State.

STATE PLAN SCHEMES

The following State Plan Schemes have been implemented during 2004-05.

Integrated Horticulture Development Scheme (IHDS)

In G.O. Ms. No.180, Agriculture (H2) Department, dated 28.5.2004, Integrated Horticulture Development Scheme was sanctioned with an outlay of ₹.766.26 lakhs. Under this scheme, planting materials, high yielding/hybrid vegetable seeds are distributed to horticultural farmers at 50 per cent subsidised cost.

Integrated Tribal Development Programme (ITDP)

In G.O. Ms. No.189, Agriculture (H2) Department, dated 28.7.2004, Integrated Tribal Development Programme was sanctioned with an outlay of ₹.39.325 lakhs. Under this scheme, horticulture inputs are distributed to Tribal farmers in Salem, Namakkal, Dharmapuri, Tiruvannamalai, Vellore, Trichy and Villupuram districts at subsidised cost. Tour-cum-Training is also given to the Tribal farmers under this scheme.

Western Ghats Development Programme (WGDP)

In G.O. Ms. No.52, Plannting, Development and Special Initiatives (TC-1) Department, dated 22.7.2004, Western Ghat Development Programme was sanctioned. Under this scheme, horticulture inputs are distributed to the horticultural farmers in Theni, Madurai, Dindigul, Virudhunagar, Coimbatore, Erode, Tirunelveli and Kanyakumari districts at 25 per cent subsidised cost.

Hill Area Development Programme (HADP)

In G.O. Ms. No.60, Planning, Development and Special Initiatives (TC 1 & 2) Department, dated 5.8.2004, Hill Area Development Programme was sanctioned with an outlay of ₹.315.00 lakhs. Under this scheme, horticultural inputs are distributed to the farmers of Nilgiris district at 25 per cent subsidised cost. The above Schemes will be continued during the year 2005-06 also. In the above schemes:-

- Focus is given for cultivation of high yielding/ hybrids and varieties, adoption of efficient water and nutrient management and Integrated Pest Management.
- Adoption of location specific and crop specific technologies is encouraged.

- Mango, Amla, Guava, Sapota and Vanilla are given importance under fruits and spices categories. High yielding/ export potential varieties are given priority.
- Under vegetables, focus is given on hybrids/ high yielding varieties which fetch good return to the farming community.
- Under Integrated Horticulture Development Scheme, horticulture plants are distributed at 50 per cent subsidy for a maximum of 1 hectare for fruits and 0.5 hectare for seeds of vegetables and flowers to each beneficiary.
- In the Tribal areas of Tamil Nadu, *viz.*, Shevroy hills, Arunoothumalai, Pachamalai, Kolli hills, Sitheri hills, Javvadu hills, Kalrayan hills, and the tribal farmers are encouraged to develop individual orchards by optimally utilizing the available water resources.
- Considering the economic status of the tribal farmers, the horticulture plants are distributed at 75 per cent subsidy under the Integrated Tribal Development Programme (ITDP).
- Watershed approach is followed in the Western Ghat Development Programme and the Hill Area Development Programme.
- Focus is given on prevention of soil erosion and preserving the Eco-System through appropriate cropping pattern.
- Planting materials and seeds are distributed at 50 per cent cost under Western Ghat Development Programme and Hill Area Development Programme.
- The State is maintaining 55 State Horticulture Farms in which Pedigree and Hybrid planting materials are produced and supplied to the farmers under various schemes.
- To update the technical knowledge of the extension officers of the Horticulture Department, in-service training is given at Central Horticulture Training Centre, Kudumianmalai in Pudukkottai district.
- A two-year diploma course is being conducted at the Horticulture Training Centre, Madhavaram (Chennai) with an intake of 40 students every year.

Promotion of Alternate Cropping in the Nilgiris

Promoting alternate crops is considered as an important measure to overcome the crisis facing the tea growers. Steps are being taken to popularize multi-tier cropping, *viz.* cultivation of silver oak, pepper, cardamom, mandarin orange, etc., along with tea plantation. Considering the prevailing agro-climatic condition in The Nilgiris, new kinds of crops like Macadamia and Peacan nuts have been identified for introduction. It is proposed to raise these crops in the State Horticultural Farms at Nanjanad and Colegrain in Nilgiris district on trial basis. Subsequently, efforts would be made to introduce these crops to the farmers in select pockets. A sum of ₹.50.00 lakhs has been sanctioned vide

G.O. Ms. No.244, Agriculture (H2) Department, dated 17.6.2004 by the Government for promoting alternate cropping in Nilgiris district and the scheme will be continued during 2005-06 also.

Tamil Nadu Precision Farming Project

As announced during last budget session, Government have sanctioned an Innovative Scheme by name "Tamil Nadu Precision Farming Project" to be implemented in Dharmapuri District and newly formed Krishnagiri district covering an area of 400 Ha. of farmers' land with Micro Irrigation and Fertigation vide G.O. Ms. No.24, Agriculture (H1) Department, dated 23.1.2004. This scheme will be implemented with a total cost of ₹.720.60 lakhs over a period of 3 years. It is programmed to cover an area of 100 Ha. in the first year, 200 Ha. in the second year and 100 Ha. in the third year at financial allocation of ₹.208.59 lakhs, 316.16 lakhs and ₹.195.85 lakhs respectively. The Tamil Nadu Agricultural University will undertake this project as a Turnkey project and implement it with the co-operation of the Departments of Horticulture, Agricultural Engineering, Agriculture, Agricultural Marketing and Agri-Business and the District Administration. High value like gherkins, hybrid tomatoes, capsicum, paprika, babycorn, white onion, bhendi, cabbage and cauliflower are proposed to be cultivated under the scheme. Under this project, 100 per cent subsidy will be given on the cost of cultivation of the first crop to the farmers selected during first year. 10 per cent of the cost of cultivation will be collected from the farmers selected during second year. 20 per cent of the cost of cultivation will be collected from the farmers selected during third year. During the year 2005-06, the above scheme will be implemented in six more districts *viz.*, Theni, Vellore, Cuddalore, Erode, Tirunelveli and Thanjavur with an financial outlay of ₹.10.00 crores. An area of 100 Ha. will be covered in each district.

SCHEMES

Under Part II Scheme a sum of ₹.120.00 lakh was sanctioned vide G.O. Ms. No.398, Agriculture (H1) Department, dated 6.8.2004 to improve the irrigation facilities in 28 State Horticulture Farms and the work is entrusted with Public Work Department and Agriculture Engineering Department and will be completed before 31.3.2005.

Tamil Nadu Horticulture Development Agency (TANHODA)

An independent nodal agency *viz.*, The Tamil Nadu Horticulture Development Agency (TANHODA) has been formed vide G.O. Ms. No.250, Agricultue (H1) Department, dated 18.6.2004 and ₹.20.00 lakhs was sanctioned under Part II 2004-05 with the following objectives.

- To promote Hi-Tech Horticulture and Precision Farming.
- To promote Public Private Partnership (PPP) for horticulture development.

- To disseminate knowledge on modern technologies in horticulture to the farming community as well as extension personnel.
- To manage production and distribution of pedigree planting materials through State Horticulture Farms.
- To implement the horticulture development schemes including the schemes under the Tamil Nadu Horticulture Development Mission and the National Horticulture Mission by accessing funds from the agencies concerned.

This agency will serve as a special purpose vehicle (SPV) for the implementation of various horticultural development programmes.

Part II Schemes (2005-06)

The following new schemes will be implemented during the year 2005-06.

- Scheme for provision of infrastructure facilities in 25 State Horticulture Farms (Farm equipments like Power Tillers with Accessories and Multipurpose Tanker) - ₹.50.00 lakhs.
- Organic farming vegetable cultivation with market linkage in 15 State Horticulture Farms - ₹.10.00 lakhs.
- Water Management - Command Area Development Programmes (Centrally sponsored scheme - shared between State and Centre)

Command Area Development Programme

The Command Area Development Programme is implemented by the Agricultural Engineering Department as a centrally sponsored and equally shared programme on 50: 50 basis between the State Government and Government of India. At present, the programme is implemented in Cauvery Command, Parambikulam Aliyar Project, Tambirabarani River Basin Project, Palar-Poranthalar Project, Krishnagiri Reservoir Project, Gadana Ramanadhi Irrigation System, Nambiyar River Basin System, Patchaiyar River Basin System, Manimuthar Irrigation System, Aanaimaduvu System, Chinnar Resevoir system and Maruthanadi System.

The programme has been completed in Lower Bhavani Project, Cumbum Valley Project, Periyar - Vaigai Project, Sathanur Reservoir Project, Amaravathy Reservoir Project, Kothayar-Chittar project and Thoppaiyar Project. The remaining projects will be covered as per the schedule approved by the Government of India. The six major components of Command Area Development Programme are as follows:

- Construction of Field Channel
- Construction of Field drain.
- Reclamation of Water logged Saline and alkaline soil areas.
- Rotational Water Supply.
- Participatory Irrigation Management
- Management Subsidy.

The Agricultural Engineering Department has also developed a water management strategy for the area below the sluice, which is crucial for agricultural production. Water management holds the key for optimisation of water use and maximisation of crop production.

Construction of Field Channel

Construction of Field channel is done by considering the needs of the head reach farmers so that their interference is eliminated and the tail end farmers also get adequate irrigation water. This brings in equity in distribution of water to all the farmers in the entire sluice command. This has been achieved by using optimum flow concept. The unit cost is ₹.10000/- per ha. in all New Project areas including Cauvery Command. For old ayacut of Cauvery Command the unit cost of construction of field channel works is ₹.4400/- per ha. An area of 9.87 lakh ha. has been covered till 31.3.2004 out of total cultivable command area of 10.97 lakh ha. with Construction of field channel component in 12 command areas. During 2004-2005, this work is being carried out in an area of 25695 ha. It is programmed to continue the scheme during 2005-2006 also.

Construction of Field Drains

Under the construction of field drains component, drains from individual fields to Government drains (or) natural drains outside the outlet are executed. The expenditure on field drains is inclusive of costs of earthwork, road cuttings and drop structures, if needed to prevent erosion. The unit cost per ha. for this component is ₹.4000/-

Reclamation of Water Logged, Saline and Alkaline Soil Areas

In this component, the area affected by Waterlogging, saline and alkaline problems within the command area are identified and measures to reclaim those patches are takenup. At present this component is taken up in Tamirabarani River Basin Project and Krishnagiri Reservoir Project at an unit cost of ₹.15000/- per ha.

Rotational Water Supply Works

Rotational Water Supply or Warabandi is a system of equitable water distribution, by turns according to a predetermined schedule specifying the day, time and duration of supply to each farmer in proportion to his holding size in an outlet command. The rotational water supply schedule is prepared by the staff of Agricultural Engineering Department after executing the on farm development works and handed over to farmers for implementation.

Rotational water supply was first introduced in Lower Bhavani Project, Periyar - Vaigai Project and Cauvery Command in the year 1984-'85. An area of 8.92 lakh ha. has been covered under rotational water supply in Tamil Nadu in 12command areas till March 2004. During 2004-2005, this work is being taken up in an area of 52750 ha. It is programmed to continue the scheme during 2005-2006 also.

Participatory Irrigation Management

To implement On Farm Development Programmes and for ensuing equity in distribution of water, farmers participation is important.

Agricultural Engineering Department is intensively encouraging itself in this work for the last 20 years.

The task of handing over the responsibilities to the farmers is being implemented. In orders to ensure farmers participation, farmers associations are formed on a three tier basis as follows.

- Farmers' Association - at sluice command level
- Farmers' Council - at distributory level
- Farmers' Federation - at project level.

Management Subsidy

In order to sustain the farmers associations/ councils already formed and for the maintenance of the on-farm infrastructures created and to continue RWS (Warabandhi) for equitable water distribution, a financial assistance is provided in the form of one time functional grant to registered Farmers' Councils at ₹.540.00 per ha. shared equally by the Central and the State Governments. To avail this, farmers have to contribute ₹.60.00 per ha. Out of the 632 Farmer's Councils registered, an amount of ₹.1478.95 lakhs has been released as management subsidy till March 2004 to 597 registered Farmers' Councils to maintain the infrastructure created under the Command Area Development Programme.

World Bank Aided Model Rehabilitation Project in Hanumanadhi Sub Basin

The World Bank aided IAIP pilot project was implemented in Hanumanadhi sub basin covering 135 ha. in the command area, which is fed by Adavinainarkoil system tanks, at a cost of ₹.81.00 lakhs. This project is construed as a threshold for a broader Model Rehabilitation Project in Hanumanadhi sub basin. The World Bank aided model Rehabilitation Project in Hanumanadhi sub basin of Tamirabarani river basin in Tirunelveli district has been successfully completed by September 30th 2004. The project has been executed and implemented in a multi disciplinary project approach involving PWD-WRO, Departments of Agricultural Engineering, Agriculture, Horticulture, Fisheries, Agricultural Marketing and TNAU. The funds of ₹.62.5 Crores was provided vide GO MS No. 33 PW (WR 1), Dated 30-1-2004, out of which ₹ 22.55 Crores was allotted to Agricultural Engineering Department. The project area encompasses Sengottai, Tenkasi and V.K. Pudur taluks. The ayacut irrigated by 50 tanks and 14 anicuts is spread over an area of 3000 ha. in 15 villages namely Ayikudi,

Elathur, Kilankadu, Kodikurichi, Kulayaneri, Kutthukalvalasai, Nainaragaram, Neduvayal, Panpoli, Rajagopalaperi, Sambavar Vadagarai, Surandai, Sundarapandiyapuram, Verakeralampudur and Vadagaraikilpidagai..An area of 2194 Ha. has been covered with an expenditure of ₹ 16.35 crores by the Agricultural Engineering Department.

Objectives

- Improving productivity of water and crop production by sharing of water and Mechanized farming.
- Uniform and optimum application of water and fertilizer by high tech micro irrigation systems. Rain water harvesting and recharge of ground water through rejuvenation of wells and farm ponds.
- Training, effecting attitudinal changes among the stakeholders and improving their socio - economic status with measures like Aquaculture in farm ponds,

High Tech Drip and Sprinkler facility

In the pre project period only two crops were cultivated in a year and most of lands were kept fallow for want of water in the third season After implementation of the project it is made possible to cultivate an additional third crop using the water conserved through micro irrigation techniques. After the project implementation an intensive cropping scheme one like below is followed.

- *Crop:* 28 Ac-inch water required crops like groundnut, vegetables.
- *Crop:* Paddy.
- *Crop:* 18 Ac-inch water requirement crops like Pulses, fodder cholam etc.

The introduction of drip and sprinkler irrigation systems in the project area on a 10 Ha block basis is estimated to save atleast 30 - 40 per cent total irrigation water. Sprinkler irrigation system are provided to the individual farmers who do not have wells and new borewells are sunk for these farmers. Drip irrigation systems are provided to the farmers who have atleast one well. Under ground PVC pipes are laid to convey water to the micro irrigation systems from the sources of water. However open field channels shall carry water during the paddy season. An area of 2194 Ha., is covered with an expenditure of ₹.1065.72 crores under high tech micro irrigation system including pipe laying.

GIS and Survey

The geographical information survey is used to get precise information on land holding, water bodies, vegetation etc., The GIS format creation in 1:1000 scale, high resolution satellite data like IRS PAN merged 1:10,000 scale, GIS in 3-D, aerial photographs are some of the advanced survey and map preparation techniques. ISDN/ Lease line connections are installed at offices of water users groups, AED office in Tenkasi, Tirunelveli and MDPPP office AED,

Commissioner of Agricultural and Commissioner of Agricultural marketing, secretariat in Chennai to provide latest information on farm practices and market trends to the farmers. These facilities have been created with ₹ 2.00 crores.

Water Harvesting Structures

Farm ponds are executed for collection of rainwater. 52 ponds have been constructed. The pond water is used for domestic, irrigation and aquaculture purposes. The defunct and low yielding wells are rejuvenated with vertical and horizontal in - well bores to improve their yield. Shallow and medium bore wells are sunk and the new bore wells are provided with energized motors. An amount of ₹ 2.84 crores has been incurred on these water-harvesting measures.

Farm Mechanization

Modern agricultural machineries like power tillers, paddy transplanters etc. have been procured at the cost of ₹.0.53 crores and are hired to the farmers.

Training to Farmers and Officers

Trainings on high tech micro irrigation systems, water-harvesting methods and farm mechanization are imparted to the water user groups and departmental officers with ₹ 0.34 crores. Documentation, video graphy and evaluation studies of the project are also included in the programme.

Farmers' Participation

The participation of the stakeholders - the farmers, is emphasized right from the decision-making and implementation stages. The farmers contributed in cash, kind and labour at various stages of the project. The progressive farmers, the women farmers and rural youth participated in every sphere of activity. The farmers will be sharing a portion of the project cost in the form of betterment levy.

Follow up on Project

With the funds from the ongoing schemes, state and other organizations like NABARD etc., the uncovered areas in this project are proposed to be covered.

JATROPHA

It has been proposed to taken up Jatropha Cultivation in 200 Ha. funded by NOVOD Board during the year 2004-05 in the following 11 State Horticulture Farms.

- G.O. Vallathirakottai, Pudukottai District 45 Ha.
- G.O. Karumanthurai, Salem District 50 Ha.
- SHF, Jeenur, Krishnagiri District 20 Ha
- SHF, Kudumianmalai, Pudukottai District 10 Ha

- SHF, Navlock, Vellore District 20 Ha
- SHF, Mulluvadi, Salem District 10 Ha
- G.O. Srivilliputhur, Virudhunagar District 12 Ha.
- SHF, Melkadirpur, Kancheepuram District 10 Ha
- SHF, Vichanthangal, Kancheepuram District 10 Ha
- SHF, Thagarakuppam, Vellore District 10 Ha
- SHF, Kannampalayam, Coimbatore District 5 Ha Total 200 ha.

NOVOD has sanctioned a sum of ₹.75.00 lakhs to the Managing Director, Tamil Nadu Watershed Development Agency, Chennai 32 towards implementation of this programme.

TARGETS AND ACHIEVEMENTS

The financial allocation for the year 2004-05 and the expenditure incurred upto February 2005 and the proposed outlay for the year 2005-2006 for the State Plan Schemes and Schemes shared between State and Centre are furnished below.

Table. Allocation for the year 2004-2005 and Proposed outlay for 2005-2006.

Unit: ₹ in lakhs

Sl. No	Scheme	Amount Sanctioned	2004 – 2005 Acht. upto Feb. 05	2005-2006 Proposed Outlay
A. State Plan Schemes				
1	Integrated Horticulture Dev. Scheme	766.260	615.080	769.400
2	Integrated Tribal Dev. Programme	39.330	35.135	41.790
3	Western Ghat Dev. Programme	135.480	105.158	132.080
4	Hill Area Development Programme	315.000	267.750	314.940
5	Precision Farming in Dharmapuri Dt.	232.950	203.000	1120.850
6	Promotion of alternate cropping(Nilgiris Dt)	50.000	48.090	50.000
7	Part-II Schemes	120.000	105.000	60.010
8	TANHODA	20.000	0	20.000
	Total	1679.020	1379.213	2509.0 0
B. Schemes shared between State and Centre				
1	Integrated Prog. for Dev. of Cashew	155.200	141.847	273.320
2	Integrated Prog. for Dev. of Fruits	218.000	186.389	318.350
3	Integrated Prog. for Dev. of Spices	60.000	47.946	180.130
4	Integrated Prog. for Dev. of Vegetables	219.000	191.644	310.000
5	Integrated Prog. for Development of Medicinal and Aromatic Plants	52.500	48.350	52.700
6	Integrated Prog. for Dev. Cocoa	12.000	11.190	12.000
7	Dev. of Commercial Floriculture	145.820	111.200	165.000
8	Development of Mushroom cultivation	10.000	9.500	10.000
9	Development of Horticulture through plasticulture interventions	14.500	13.030	14.500
10	Innovative Programmes for			

Development of Horticulture	113.900	43.455	90.000
Total	1000.920	804.471	1426.000
Grand Total (A+B)	2679.940	2183.684	3935.070

CONCLUSION

Tamil Nadu is one of the leading horticulture States in India contributing 7.7 per cent to the National Horticultural production with 5.7 per cent of the national level area. Tamil Nadu has been blessed with diversified agro-climatic conditions, suitable for a wide range of horticulture crops like fruits, vegetables, spices, plantation crops, flowers and medicinal plants. A large extent of wastelands and under-utilized lands are available in the State for horticulture development. Tamil Nadu has a long coastal belt of 1000 km. suitable for crops like cashew, coconut, tropical orchids etc. The southern part of Tamil Nadu has the potential for growing off-season mangoes and grapes. A lot of awareness has been created among the farmers of Tamil Nadu about cultivation of high value horticulture crops. It is aimed to achieve 8 per cent annual growth rate during X Five-Year Plan in the horticultural sector. The area coverage in 2003-04 was 8.24 lakh hectares, a little less than the normal area of 8.52 lakh hectares.

During 2004-05, despite the failure of the monsoon, the area coverage would still be around 8.91lakh hectares. In the year 2005-06, an area of 9.73 lakh hectares is expected to be covered.

6

Horticulture Pest in Gardens and Landscapes Management

SCALES: PESTS IN GARDENS AND

Scales are sucking insects that insert their tiny, strawlike mouthparts into bark, fruit, or leaves, mostly on trees and shrubs and other perennial plants. Some scales can seriously damage their host, while other species do no apparent damage to plants even when scales are very abundant. The presence of scales can be easily overlooked, in part because they do not resemble most other insects.

IDENTIFICATION

Adult female scales and immatures (nymphs) of most species are circular to oval, wingless, and lack a separate head or other easily recognizable body parts. Some scales change greatly in appearance as they grow, and some species have males and females that differ in shape, size, and Colour. Adult males are rarely seen and are tiny, delicate, white to yellow insects with one pair of wings and a pair of long antennae. Some scale species lack males and the females reproduce without mating.Armored scales and soft scales are the most common types (or families). Scales in other families include important pests of cactus, elm, oak, sycamore, and various conifers. Common scales and their tree and shrub.

Cottony cushion scale, European elm scale, soft scales, and certain other scales secrete sticky honeydew. Armored scales, oak pit scales, and sycamore scale do not excrete honeydew. It is important to correctly distinguish the scale family (*e.g.*, armored versus soft scale) and often the particular species of scale to determine whether control is warranted; and if so, what methods and timing of control action are effective. For example, sago palms can be infested by the similar-looking cycad scale and oleander scale. Even very high populations of oleander scale are harmless to most plants, but cycad scale warrants control because it causes serious damage and can kill sago palms. Insecticides differ in their effectiveness for certain scale types. imidacloprid, a popular systemic insecticide, controls soft scales and certain other scales but does not control armored scales or cottony cushion scale.

Armored Scales

Armored scales, family Diaspididae, have a flattened, platelike cover that is less than 1/8 inch in diameter. The covers often have a differently colored, slight protuberance (exuviae or "nipple") and concentric rings may form as nymphs (immatures) grow and their cover enlarges. The actual insect body is underneath the cover; if you remove the cover, the insect body will remain on the plant. Armored scales do not produce honeydew. Damaging species include cycad scale, euonymus scale, oystershell scale, and San Jose scale.

Soft Scales

Soft scales, family Coccidae, grow up to 1D4 inch long and have a smooth, cottony, or waxy surface. At maturity, soft scales are usually larger and more rounded and convex (humped) than armored scales. Their surface is the actual body wall of the insect and cannot be removed; flipping the cover removes the insect body and cover together. Soft scales and certain other types feed on phloem sap and excrete abundant, sticky honeydew, which drips on plants and surfaces underneath and promotes the growth of blackish sooty mold. Soft scales include black scale, brown soft scale, Kuno scale, lecanium scales, and tuliptree scale.

Scale Look-Alikes

Various other organisms resemble scales but have different biology and management. These include California laurel aphid *(Euthoracaphis umbellulariae)*, coconut mealybug *(Nipaecoccus nipae)*, cypress bark mealybug*(Ehrhornia cupressi)*, palm aphid *(Cerataphis brasiliensis*), whitefly nymphs, and psyllids, such as lemongum lerp psyllid *(Cryptoneossa triangula)* and redgum lerp psyllid *(Glycaspis brimblecombei*). Diamond "scale," which infests palms, is actually the fruiting bodies of a blackish fungus (*Phaeochoropsis neowashingtoniae)*.

LIFE CYCLE

Scales hatch from an egg and typically develop through two nymphal instars (growth stages) before maturing into an adult. Each instar can change greatly as it ages, so many scales appear to have more than two growth stages. At maturity, adult females produce eggs that are usually hidden under their bodies, although some species secrete their eggs externally under prominent cottony or waxy covers. Eggs hatch into tiny crawlers (mobile first instar nymphs), which are yellow to orangish in most species. Crawlers walk over the plant surface, are moved to other plants by wind, or are inadvertently transported by people or birds. Crawlers settle down and begin feeding within a day or two after emergence.

Settled nymphs may spend their entire life in the same spot without moving as they mature into adults. Nymphs of some species can move slowly, such as soft scales that feed on deciduous hosts and move from foliage to bark in the fall before leaves drop. For species with multiple generations, all scale life stages may be present throughout the year in areas with mild winters.

Armored Scales

Most species of armored scales have several generations a year and overwinter primarily as first instar nymphs and adult females. Except for crawlers and adult males, armored scales lack obvious appendages and spend their entire life feeding at the same spot.

Soft Scales

Most soft scales have one generation each year and overwinter as second instar nymphs. The brown soft scale is an exception; it has multiple generations and females and nymphs can be present throughout the year. Most immature soft scales retain their barely visible legs and antennae after settling and are able to move, although slowly.

DAMAGE

Some scale species, when abundant, weaken a plant and cause it to grow slowly. Infested plants appear water stressed, leaves turn yellow and may drop prematurely, and plant parts that remain heavily infested may die. The dead brownish leaves may remain on scale-killed branches, giving plants a scorched appearance. If the scale produces honeydew, this sticky excrement, sooty mold, and the ants attracted to honeydew can annoy people even when scales are not harming the plant.

The importance of infestations depends on the scale species, the plant species and cultivar, environmental factors, and natural enemies. Populations of some scales can increase dramatically within a few months when the weather is warm, and honeydew-seeking ants protect scales from their natural enemies. Plants are not harmed by a few scales and even high populations of certain species apparently do not damage plants.

MANAGEMENT

Many species are usually well controlled by beneficial predators and parasites (natural enemies). Exceptions are when natural enemies are disrupted by ants, dust, or the application of persistent broad-spectrum insecticides. Preserving (conserving) parasites and predators (such as by controlling pest-tending ants) may be enough to bring about gradual control of certain scales as natural enemies become more abundant.

A well-timed and thorough spray of horticultural (narrow-range) oil during the dormant season, or soon after scale crawlers are active in late winter to early summer, can provide good control of most species of scale. Certain scale problems on large plants and hosts especially sensitive to scale damage may warrant the application of a systemic insecticide. If plants perform poorly or are repeatedly damaged by pests, the best course of action may be to replace the plant with a pest-resistant species or cultivar that is better adapted to the site conditions.

Monitoring

Periodically check to ensure that plants have a good growing environment and are receiving appropriate cultural care. Inspect plants to determine whether female scales, nymphs, honeydew, sooty mold, or ants and other pests are present. Before applying insecticide, examine a portion of the scales to determine whether they are dead or parasitized as described below; for example, they fail to exude fluid when squished. If a large proportion of scales are dead or parasitized by natural enemies, consider delaying a treatment decision and monitor the population again later before deciding whether to apply pesticide. Tape traps for crawlers and honeydew monitoring are useful in certain situations for determining the need and best timing for pesticide application.

Inspect trunks for ants periodically during the growing season. If the descending ants have swollen, almost translucent abdomens, they may be feeding on honeydew produced by scales or other insects. Trace back trail-making ants to locate colonies of the honeydew-producing insects.

There are no quantitative action guidelines for deciding whether pesticide application for scales is warranted. Monitor and record scale densities and use the density that caused damage (dieback or unacceptable honeydew) as your preliminary control action threshold. As you gain experience, refine this threshold over time for your local situation.

Tape Traps

Transparent double-sided sticky tape can be used to effectively time a foliar insecticide application. During the spring before crawlers begin to emerge, tightly encircle each of several scale-infested twigs or branches with transparent tape that is sticky on both sides, available at fabric or craft stores. Double over the loose end of the tape several times to make it easier to remove. Place a tag or flagging near each tape so you can readily find it. Change the tapes at weekly intervals. After removing the old tape, wrap the twig at the same location with fresh tape. Preserve the old sticky tapes by sandwiching them between a sheet of white paper and clear plastic. Label the tapes with the date, location, and host plant from which they were collected. Scale crawlers get stuck on the tape and appear as yellow or orange specks. Examine the tape with a hand lens to distinguish the crawlers (which are round or oblong and have very short appendages) from pollen and dust. Use a hand lens to examine the crawlers beneath mature female scales on bark or foliage to be certain of crawler appearance. Other tiny creatures, including mites, may also be caught in the tape.

Visually compare the tapes collected on each sample date. If a spring or summer foliar insecticide application is planned, unless another time is recommended for that species, spray after crawler production (abundance in traps) has peaked and definitely begun to decline, which is soon after most crawlers have settled.

Honeydew Monitoring

Honeydew drippings from plants can be efficiently monitored using water-sensitive paper, which is commonly used for monitoring insecticide droplets and calibrating sprayers. Products include bright yellow cards that produce distinct blue dots upon contact with honeydew or water. Regularly monitoring honeydew beneath plants, such as the number of drops during four hours on the same time of day once a week, can help to develop thresholds and evaluate effectiveness of the treatment. Honeydew monitoring is useful where there is a low tolerance for dripping honeydew, when managing many trees, such as along city streets or in parks, and on tall trees where the honeydew-producing insects may be located too high to easily observe.

Cultural Control

Provide plants with good growing conditions and proper cultural care, especially appropriate irrigation, so they are more resistant to scale damage. You can prune off heavily infested twigs and branches, if they are limited to a few parts of small plants. In areas with hot summers, pruning to open up canopies can reduce populations of black scale, citricola scale, cottony cushion scale, and possibly other scales by increasing scale mortality from exposure to heat and parasites. Consider replacing problem-prone plants.

Biological Control

Scales are preyed upon by small parasitic wasps and many predators, including certain beetles, bugs, lacewings, and mites. Predatory lady beetle (ladybug) species of *Chilocorus*, *Hyperaspis*, and *Rhyzobius*, can easily be overlooked because the adults of many species are tiny or colored and shaped like scales, and their larvae may feed hidden beneath scales. *Hyperaspis* species are tiny, shiny, black lady beetles with several red, orange, or yellow spots on the back. *Rhyzobius lophanthae* has a reddish head and underside and a grayish back densely covered with tiny hairs. Thetwicestabbed lady beetle, *Chilocorus orbus*, is shiny black with two red spots on its back.

Often the most important natural enemies of scales are parasitic wasps, including species of *Aphytis*, *Coccophagus*, *Encarsia*, and *Metaphycus*. The female wasp lays one or several eggs in or on each scale, where the tiny maggotlike wasp larvae feed. Parasitized scales may become puffy or darken in comparison with unparasitized scales. Sometimes the immature parasites are visible through the scale surface. After completing the larval stage and pupating, the emerging adult of internal parasites typically leaves a round exit hole in the scale it killed. With external parasites that feed outside the scale body, but under the cover of armored scales, their maggotlike larvae may be observed by prying off the scale cover. Natural enemies are commercially available for release against California red scale and perhaps certain other scales. However, conserving resident natural enemies is a more efficient and longer lasting strategy than buying and releasing beneficials in gardens and landscapes.

Ant control, habitat manipulation, and pesticide management are the key natural enemy conservation strategies. If ants are abundant, selectively control them. Grow a variety of flowering plants to help attract and support natural enemies. Adults of predatory bugs, lacewings, lady beetles, and parasitic wasps live longer, lay more eggs, and kill more scales when they have plant nectar or pollen and insect honeydew to feed on. Avoid creating dust because it interferes with natural enemies. For example, rinse small plants when foliage becomes dusty. Depending on the scale species and the extent to which biological control has been disrupted, it will take several months of conservation efforts (such as controlling ants and dust and avoiding application of persistent insecticides) or until the next season or longer before scale populations are reduced by biological control. If current levels of scales are intolerable, spray insecticidal oils to reduce scale populations while conserving natural enemies.

ANT CONTROL

Because ants attack and feed on scale parasites and predators, control ants if they are tending scales. To deny ants access to plant canopies, prune branches or weeds that provide a bridge between buildings or the ground and apply a sticky material (Tanglefoot) to trunks. Wrap the trunk with a collar of fabric tree wrap, heavy paper, or masking tape to avoid injury to bark; wedge pliable wrap snugly into cracks and crevices; and coat the wrap with the sticky material.

A barrier band about 2 to 6 inches wide should be adequate in most situations. Inspect wraps at least several times a year for damage to bark and remove and relocate any wrap at least once a year to minimize bark injury. Periodically stir sticky material with a stick to prevent ants from crossing on collected debris. Avoid applying sticky material to horizontal surfaces where birds may roost.

Place enclosed pesticide baits (insecticide mixed with an attractant) near nests or on ant trails beneath plants. Effective, slow-acting bait insecticides work over a period of days so that before ants die they will spread the toxicant among many other ants during food sharing. Boric acid, fipronil, and hydramethylnon are examples of insecticides used in ant baits. Although baits require users to be patient, they can be much more effective than sprays. Sprays only kill foraging workers, while ant baits are carried back to the nests where reproductive queens and the entire colony underground can be killed.

CHEMICAL CONTROL

Before applying insecticide, make sure plants are receiving appropriate cultural care and take steps to conserve natural enemies. Check a portion of the scales to be certain they are alive and to evaluate the extent of parasitism. To know how and when to effectively make an application, learn more about the available insecticides and the biology of your pest species. Completely read and follow the product label instructions for the safe and effective use of the insecticide. Insecticides can have unintended effects, such as contaminating water, poisoning natural enemies and pollinators, and causing secondary pest outbreaks.

NON-RESIDUAL, CONTACT INSECTICIDES

Where plants can be sprayed, complete spray coverage of infested plant parts with horticultural oil at the proper time provides good control of most scales. Horticultural oils (*e.g.*, Bonide Horticultural Oil and Monterey Horticultural Oil) are specially refined petroleum products, often called narrow-range, superior, or supreme oils. Other Non-persistent, contact sprays for garden and landscape plants include insecticidal soap (Safer Brand Insect Killing Soap Concentrate II), neem oil (Bayer Advanced Natria Neem Oil Concentrate, Green Light Neem, Garden Safe Brand Neem), canola oil (Bayer Advanced Natria Multi-Insect Control), and other botanical (plant-derived) oils.

These insecticides have low toxicity to people and pets and relatively little adverse impact on the populations of pollinators and natural enemies and the benefits they provide. To obtain adequate control, thoroughly wet the infested plant parts with spray, typically shoot terminals and the underside of leaves. More than one application per growing season may be needed, especially if the targeted pest has more than one generation a year. Thorough spray coverage is especially critical when treating armored scales and oak pit scales as these scales are generally less susceptible to pesticides than soft scales.

Dormant Season

To control most scales overwintering on deciduous woody plants, thoroughly spray the bark of terminal shoots with oil during winter. For oak pit scales, sycamore scale, and other harder to control species, spray during the plant's delayed-dormant period, which is after the buds swell but before buds open. Do not spray oystershell or olive scales during the dormant season because susceptible stages of these species are not present during winter.

Foliage Spray

Horticultural oil is effective in spring or summer on deciduous plants when sprayed soon after most crawlers have emerged and most scales are in the young nymph stage. Late spring and summer are also the times to spray avocado, citrus, and many other broadleaf evergreens. Thoroughly cover with spray the plant parts where scales occur, typically on twig terminals and the underside of leaves.

Precautions on Using Oils

Follow product labels, which may say to not spray certain plant species or mix oil with certain other products. For example, oil will remove the desirable bluish tinge from blue spruce foliage, although the plant's health is not impaired. Do not mix oil with chlorothalonil, sulfur, and certain other fungicides; and do not apply oil within 3 weeks of an application of sulfur-containing compounds, such as wettable sulfur. Do not apply oil or other insecticides when it is foggy, freezing (under 32°F), hot (over 90°F), when relative humidity is above 90%, or if rain is expected in the next 24 hours. Especially at locations with hot weather, be sure plants are well irrigated before spraying foliage.

SYSTEMIC INSECTICIDES

Systemic insecticides are absorbed by one plant part (*e.g.*, trunks or roots) and moved (translocated) to leaves and other plant parts. In comparison with systemics that are sprayed onto foliage, products labeled for soil drench or injection, or for trunk injection or spray minimize environmental contamination and may be more effective than contact insecticides. Trunk application of an effective systemic insecticide can provide relatively rapid control. There is a longer time delay between soil application and insecticide action. Some uses require hiring a professional pesticide applicator. Certain home-use products can easily be drenched into soil around the tree trunk using the mix-and-pour method.

Systemic insecticides for use on landscape plants include Neo-nicotinoids (acetamiprid, dinotefuran, imidacloprid, and thiamethoxam) and the organophosphate acephate (Lilly Miller Ready-to-Use Systemic, Orthene). Properly applied, one application of an effective product may provide season-long control. However, Neo-nicotinoids vary in effectiveness for scale control. For example, acetamiprid (Ortho Flower, Fruit and Vegetable Insect Killer) controls soft scales but is not very effective on armored scales and can only be applied by spraying foliage. Imidacloprid controls European elm scale and most soft scales but does not control cottony cushion scale and most armored scales.

Dinotefuran (Green Light Tree and Shrub Insect Control with Safari 2G, Safari) controls most types of scales. Some of these products are for licensed professional applicators only.

Some systemic insecticides can cause spider mite outbreaks. Foliage sprays of systemics can be toxic to beneficial insects that contact spray or treated leaves. Systemics can translocate into flowers and have adverse effects on natural enemies and pollinators that feed on nectar and pollen. Do not apply systemic insecticides to plants during flowering or shortly before flowering; wait until after plants have completed their seasonal flowering unless the product's label directions say otherwise. With foliage spraying and soil application, when possible, wait until nearby plants also have completed flowering as spray can drift onto nearby plants or their roots may take up some of the soil-applied insecticide.

When applying systemic insecticide, use soil application or a trunk spray whenever possible. With trunk injection and implantation it is difficult to repeatedly place insecticide at the proper depth. Trunk injection and implantation also injure woody plants and can spread plant pathogens on contaminated tools. When injecting or implanting into multiple plants, scrub any plant sap from tools or equipment that penetrate bark and disinfect tools with a registered disinfectant (*e.g.*, bleach) before moving to work on each new plant. At least one to two minutes of disinfectant contact time between contaminated uses is generally required. Consider rotating work among several tools and using a freshly disinfected tool while the most recently used tools are being soaked in disinfectant. Avoid methods that cause large wounds, such as implants placed in holes drilled in trunks. Do not implant or inject into roots or trunks more than once a year.

APHIDS PESTS IN GARDENS AND LANDSCAPES

Aphids are small, soft-bodied insects with long slender mouthparts that they use to pierce stems, leaves, and other tender plant parts and suck out fluids. Almost every plant has one or more aphid species that occasionally feed on it. Many aphid species are difficult to distinguish from one another; however, management of most aphid species is similar.

IDENTIFICATION

Aphids have soft pear-shaped bodies with long legs and antennae and may be green, yellow, brown, red, or black depending on the species and the plants they feed on. A few species appear waxy or woolly due to the secretion of a waxy white or gray substance over their body surface. Most species have a pair of tubelike structures calledcornicles projecting backward out of the hind end of their body. The presence of cornicles distinguishes aphids from all other insects.

Generally adult aphids are wingless, but most species also occur in winged forms, especially when populations are high or during spring and fall. The ability to produce winged individuals provides the pest with a way to disperse to other plants when the quality of the food source deteriorates.

Although they may be found singly, aphids often feed in dense groups on leaves or stems. Unlike leafhoppers,plant bugs, and certain other insects that might be confused with them, most aphids don't move rapidly when disturbed.

LIFE CYCLE

Aphids have many generations a year. Most aphids in California's mild climate reproduce asexually throughout most or all of the year with adult females giving birth to live offspring—often as many as 12 per day—without mating. Young aphids are called nymphs. They molt, shedding their skin about four times before becoming adults. There is no pupal stage. Some species produce sexual forms that mate and produce eggs in fall or winter, providing a more hardy stage to survive harsh weather and the absence of foliage on deciduous plants. In some cases, aphids lay these eggs on an alternative host, usually a perennial plant, for winter survival. When the weather is warm, many species of aphids can develop from newborn nymph to reproducing adult in seven to eight days. Because each adult aphid can produce up to 80 offspring in a matter of a week, aphid populations can increase with great speed.

DAMAGE

Low to moderate numbers of leaf-feeding aphids aren't usually damaging in gardens or on trees. However, large populations can turn leaves yellow and stunt shoots; aphids can also produce large quantities of a sticky exudate known as honeydew, which often turns black with the growth of a sooty mold fungus. Some aphid species inject a toxin into plants, which causes leaves to curl and further distorts growth. A few species cause gall formations.

Aphids may transmit viruses from plant to plant on certain vegetable and ornamental plants. Squash, cucumber, pumpkin, melon, bean, potato, lettuce, beet, chard, and bok choy are crops that often have aphid-transmitted viruses associated with them.

The viruses mottle, yellow, or curl leaves and stunt plant growth. Although losses can be great, they are difficult to prevent by controlling aphids, because infection occurs even when aphid numbers are very low; it takes only a few minutes for the aphid to transmit the virus, while it takes a much longer time to kill the aphid with an insecticide.

A few aphid species attack parts of plants other than leaves and shoots. The lettuce root aphid is a soil dweller that attacks lettuce roots in spring and summer, causing lettuce plants to wilt and occasionally die. In fall, this species often moves to poplar trees, where it overwinters in the egg stage and produces leaf galls in spring. Thewoolly apple aphid infests woody parts of apple roots and limbs, often near pruning wounds, and can cause overall tree decline if roots are infested for several years. Heavy infestations of crown and root aphids on carrots may weaken tops, causing them to tear off when carrots are harvested.

MANAGEMENT

Although aphids seldom kill a mature plant, the damage they do and unsightly honeydew they generate sometimes warrant control. Consider the Non-chemical controls, as most insecticides will destroy beneficial insects along with the pest. On mature trees, such as in citrus orchards, aphids and the honeydew they produce can provide a valuable food source for beneficial insects.

Monitoring

Check your plants regularly for aphids—at least twice a week when plants are growing rapidly—in order to catch infestations early, so you can knock or hose them off or prune them out. Many species of aphids cause the greatest damage in late spring when temperatures are warm but not hot (65°-80°F). For aphids that cause leaves to curl, once aphid numbers are high and they have begun to distort leaves, it's often difficult to control these pests, because the curled leaves shelter aphids from insecticides and natural enemies.

Aphids tend to be most prevalent along the upwind edge of the garden and close to other infested plants of the same species, so make a special effort to check these areas. Many aphid species prefer the underside of leaves, so turn leaves over when checking for aphids. On trees, clip off leaves from several areas of the tree. Also check for evidence of natural enemies such as lady beetles, lacewings, syrphid fly larvae, and the mummified skins of parasitized aphids. Look for disease-killed aphids as well; they may appear off color, bloated, flattened, or fuzzy. Substantial numbers of any of these natural control factors can mean the aphid population may be reduced rapidly without the need for treatment.

Ants are often associated with aphid populations, especially on trees and shrubs, and frequently are a clue that an aphid infestation is present. If you see large numbers of ants climbing your tree trunks, check higher up the tree for aphids or other honeydew-producing insects that might be on limbs and leaves. To protect their food source, ants ward off many predators and parasites of aphids. Managing ants is a key component of aphid management.

In landscape settings, you can monitor aphids by using water-sensitive paper to measure honeydew dripping from a tree. This type of monitoring is of particular interest where there is a low tolerance for dripping honeydew, such as in groups of trees along city streets or in parks and for tall trees where aphid colonies may be located too high to detect.

Biological Control

Natural enemies can be very important for controlling aphids, especially in gardens not sprayed with broad-spectrum pesticides (*e.g.*, organophosphates, carbamates, and pyrethroids) that kill natural enemy species as well as pests. Usually natural enemy populations don't appear in significant numbers until aphids begin to be numerous.

Among the most important natural enemies are various species of parasitic wasps that lay their eggs inside aphids. The skin of the parasitized aphid turns crusty and golden brown, a form called a mummy. The generation time of most parasites is quite short when the weather is warm, so once you begin to see mummies on your plants, the aphid population is likely to be reduced substantially within a week or two.

Many predators also feed on aphids. The most well known are lady beetle adults and larvae, lacewing larvae, soldier beetles, and syrphid fly larvae. Naturally occurring predators work best, especially in garden and landscape situations.

Lady Beetle Releases

Applying commercially available lady beetles (the convergent lady beetle, *Hippodamia convergens*) may give some temporary control when properly handled, although most of them will disperse from your yard within a few days. If releasing lady beetles, keep them refrigerated until just before letting them go, doing so at dusk, as those released in broad daylight will fly away immediately. Mist the lady beetles with water just before release, and also mist the surface of the plant you are releasing them onto. Place the lady beetles at the base of infested plants or in the crotches of low branches. Lady beetles will crawl higher into the plant in search of aphids. University of California research indicates that high numbers of lady beetles are required to control aphids. One large, heavily infested rose bush required two applications, spaced a week apart, of about 1,500 lady beetles each. Aphids are very susceptible to fungal diseases when it is humid. These pathogens can kill entire colonies of aphids when conditions are right. Look for dead aphids that have turned reddish or brown; they'll have a fuzzy, shriveled texture unlike the shiny, bloated, tan-colored mummies that form when aphids are parasitized. Weather can also impact aphids. Summer heat in the Central Valley and desert areas reduces the populations of many species, and aphid activity is also limited during the coldest part of the year. However, some aphids may be active year-round, especially in the milder, central coastal areas of California.

Ant Management

In some situations ants tend aphids and feed on the honeydew aphids excrete. At the same time, ants protect the aphids from natural enemies. If you see ants crawling up aphid-infested trees or woody plants, put a band of sticky material (*e.g.*, Tanglefoot) around the trunk to prevent ants from climbing up. (Don't apply sticky material directly to the bark of young or thin-barked trees or to trees that have been severely pruned, as the material may have phytotoxic effects. Wrap the trunk with fabric tree wrap or duct tape and apply sticky material to the wrap.) Alternatively, ant stakes or containerized baits may be used on the ground to control ants without affecting aphids or their natural enemies. Prune out other ant routes such as branches touching buildings, the ground, or other trees.

CULTURAL CONTROL

Before planting vegetables, check surrounding areas for sources of aphids and remove these sources. Some aphids build up on weeds such assowthistle and mustards, moving onto related crop seedlings after they emerge. On the other hand, these aphid-infested weeds can sometimes provide an early source of aphid natural enemies. Always check transplants for aphids and remove them before planting. Where aphid populations are localized on a few curled leaves or new shoots, the best control may be to prune out these areas and dispose of them. In large trees, some aphids thrive in the dense inner canopy; pruning out these areas can make the habitat less suitable.

High levels of nitrogen fertilizer Favour aphid reproduction, so never use more nitrogen than necessary. Instead, use a less soluble form of nitrogen and apply it in small portions throughout the season rather than all at once. Slow-release fertilizers such as organic fertilizers or urea-based time-release formulations are best.

Because many vegetables are susceptible to serious aphid damage primarily during the seedling stage, reduce losses by growing seedlings underprotective covers in the garden, in a greenhouse, or inside and then transplanting them when the seedlings are older and more tolerant of aphid feeding. Protective covers will also prevent transmission of aphid-borne viruses.

Silver-colored reflective mulches have been successfully used to reduce transmission of aphid-borne viruses in summer squash, melon, and other susceptible vegetables. These mulches repel invading aphid populations, reducing their numbers on seedlings and small plants. Another benefit is that yields of vegetables grown on reflective mulches are usually increased by the greater amount of solar energy reflecting onto leaves. To put a reflective mulch in your garden, remove all weeds and cover beds with mulch. Bury the edges with soil to hold them down. After the mulch is in place, cut or burn 3- to 4-inch diameter holes and plant several seeds or a single transplant in each one. In addition to repelling aphids, leafhoppers, and some other insects, the mulch will enhance crop growth and control weeds. When summertime temperatures get high, however, remove mulches to prevent overheating plants.

Ready-to-use reflective mulch products include silver-colored plastic sold in rolls. You can also make your own by spray-painting construction paper, landscape fabric, or clear plastic. If you use plastic mulches, you will need to use drip irrigation underneath. Landscape fabric and most paper mulches will allow water to flow through.

Another way to reduce aphid populations on sturdy plants is to knock off the insects with a strong spray of water. Most dislodged aphids won't be able to return to the plant, and their honeydew will be washed off as well. Using water sprays early in the day allows plants to dry off rapidly in the sun and be less susceptible to fungal diseases.

CHEMICAL CONTROL

When considering whether to apply insecticides for aphid control, remember that most larger plants can tolerate light to moderate levels of aphids with little damage. Larger aphid populations often rapidly decline due to biological control or when hot temperatures arrive. Often a forceful spray of water or water-soap solution, even on large street trees, when applied with appropriate equipment, will provide sufficient control.

If insecticides are needed, insecticidal soaps and oils are the best choices for most situations. Oils may include petroleum-based horticultural oils or plant-derived oils such as neem or canola oil. These products kill primarily by smothering the aphid, so thorough coverage of infested foliage is required. Apply these materials with a high volume of water, usually a 1 to 2% oil solution in water, and target the underside of leaves as well as the top. Soaps, neem oil, and horticultural oil kill only aphids present on the day they are sprayed, so applications may need to be repeated. Although these materials can kill some natural enemies that are present on the plant and hit by the spray, they leave no toxic residue so they don't kill natural enemies that migrate in after the spray.

These and other insecticides with contact-only activity are generally ineffective in preventing damage from aphids such as the leaf curl plum aphid or the woolly ash aphid, which are protected by galls or distorted foliage. Also, don't use soaps or oils on water-stressed plants or when the temperature exceeds 90°F. These materials may be phytotoxic to some plants, so check labels and test the materials on a portion of the foliage several days before applying a full treatment.

Supreme- or superior-type oils will kill overwintering eggs of aphids on fruit trees if applied as a delayed-dormant application just as eggs are beginning to hatch in early spring. (On plums dormant applications right after leaves have fallen in early November are preferred.) These treatments won't give complete control of aphids and probably aren't justified for aphid control alone but will also control soft scale insects if they are a problem. Common aphid species controlled with these types of oils include the woolly apple aphid, green apple aphid, rosy apple aphid, mealy plum aphid, and black cherry aphid.

Many other insecticides are available to control aphids in the home garden and landscape, including foliar-applied formulations of malathion, permethrin, and acephate (Non-food crops only). While these materials may kill higher numbers of aphids than soaps and oils, their use should be limited, because they also kill the natural enemies that provide long-term control of aphids and other pests, and they are associated with bee kills and environmental problems. Repeated applications of these materials may also result in resistance to the material.

Insecticides such as oils and soaps are also safer to use when children and pets may be present. Formulations combining insecticidal soaps and pyrethrins may provide slightly more knockdown than soaps alone yet have fewer negative impacts on natural enemies than malathion, permethrin, and acephate, because pyrethrins break down very quickly.

Systemic insecticides are also available for aphid management, primarily for woody ornamentals. These materials, including imidacloprid, are very effective and are especially useful for serious infestations of aphids such as the woolly hackberry aphid, which is often not effectively controlled by biological control or less toxic insecticides. Imidacloprid can have negative impacts on predators, parasitoids, and pollinators, so its use should be avoided where soaps and oils will provide adequate control.

To protect pollinators, don't apply imidacloprid or other systemic insecticides to plants in bloom or prior to bloom. Home-use soil-applied imidacloprid products are often diluted with water in a bucket and poured around the base of the tree or plant. Professional applicators can use soil injectors, which provide better control with less Run-off potential. Applications are usually made in spring when aphids first become apparent. Adequate rain or irrigation is required to move the product through the soil to the roots and up into large trees, and it may take several weeks to see an effect on aphids feeding on leaves. One application on hackberry is enough to control hackberry woolly aphid for two to three years.

A STRATEGY FOR PEST CONTROL ON LANDSCAPE PLANTS

The diversity of plant materials in the residential landscape translates to a wide range of potential pest problems. Keeping plants healthy can prevent many of them. Plants that are stressed from undesirable cultural or site conditions grow poorly and become prime targets for pests. For example, a landscape plant is more likely to have spider mite problems when it is planted in a hot, dry area. Likewise, plants that have been damaged by a lawn mower are more likely to have borer problems than undamaged, healthy ones. The best way to discourage plant pest problems is to fulfill their cultural requirements.

CONTROL OF COMMON PESTS OF LANDSCAPE PLANTS

Many species of insects or mites attack Georgia landscape plants. Homeowners have difficulty controlling these pests because they often are not aware of the problem until both the infestation and the damage are extensive, or they may apply insecticides improperly or at the wrong time. An effective pest control Programme on landscape plants begins with proper plant selection. Select plants that are less prone to insect problems when designing, purchasing or replacing landscape plants. Secondly, on a weekly basis during the growing season, closely inspect landscape plants for developing pest populations. Remember that some insects and mites are minute in size and are frequently found on the undersides of leaves. A close and thorough inspection is necessary to detect these pests. Thirdly, if you find a developing pest population that you cannot identify, get some assistance with the identification from your local county extension office or a reliable nursery. You must know what pests you are trying to control. Brief descriptions of major groups of pests (or the damage they cause) found on Georgia landscape plants are provided to assist with identification.

Sucking Pests

Sucking insect pests cause damage by removing sap from plant tissues. Symptoms of infestation include wilting of plant tissues, stunting, curling or distortion of new growth, chlorotic spots or stippling of leaf surface, or a sticky substance or black fungal growth on the upper leaf surface. Common insects and mites causing this type damage include aphids, scale insects, mealybugs, lace bugs, whiteflies, and spider mites.

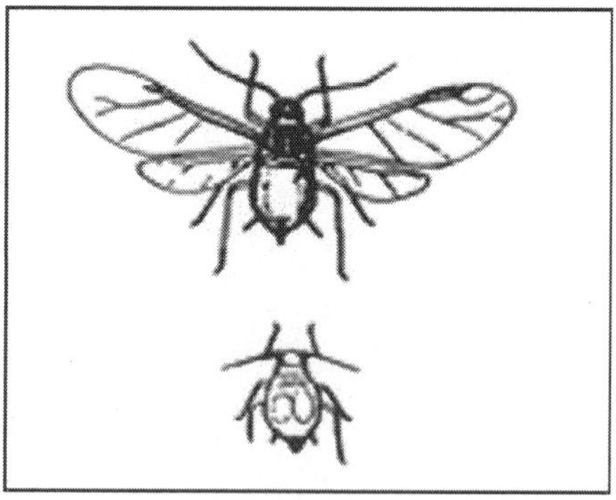

Fig. Aphids.

Aphids are small (about 1/4 inch in length), soft-bodied insects that vary in Colour from green to yellow to black. Some species are winged during certain times of the year. Generally, aphids can be recognized by their cornicles, a pair of tube-like structures projecting from the rear of their bodies. They are frequently found in large numbers clustered together on the backs of leaves or on the stems of new growth. *Aphids: adult winged female (top); adult wingless female (bottom)*

Scale Insects and Mealybugs

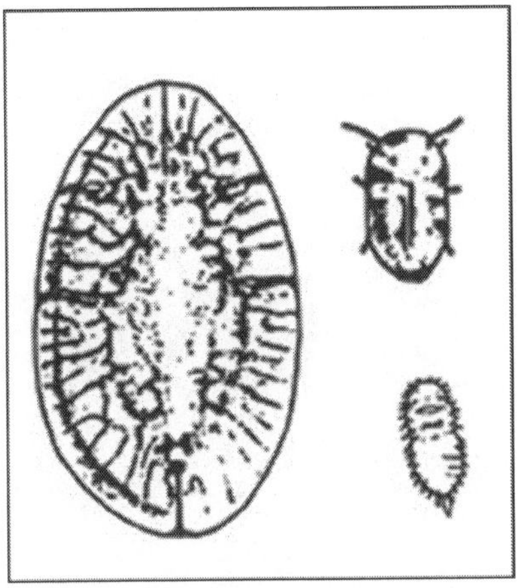

Scale insects are very small, soft-bodied pests that secrete a protective covering over their bodies. These coverings vary in Colour from white to red to black. Some are flattened while others are more turtle-shaped. This covering protects the scale and makes control difficult. Scale insects are most easily controlled when insecticide applications are timed for the firstinstars or "crawler" stage of the scale. Mealybugs are very similar to scale insects, however they secrete a white waxy material over their bodies.

Mealybugs also move about on the host plant to feed. Mealybugs are most susceptible to insecticide applications when they are young and have not formed a thick covering over their bodies.

L-R: Adult brown soft scale; scale *"crawler" (top);* mealybug *(bottom)*

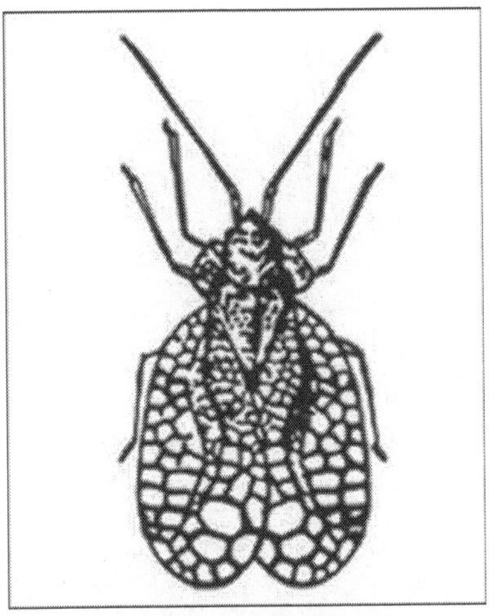

Fig. Lace Bugs.

Lace bugs get their name from the appearance of the area behind their head and the wing covers. The area forms a lacelike covering over the body of the insect. They are 1/8 to 1/4 inch in length and are partially transparent. Lace bug damage appears on the upper leaf surface as white to yellow chlorotic spots. The lower leaf surfaces will be cluttered with black spots and the old cast skins of immature lace bugs.

Adult lace bug

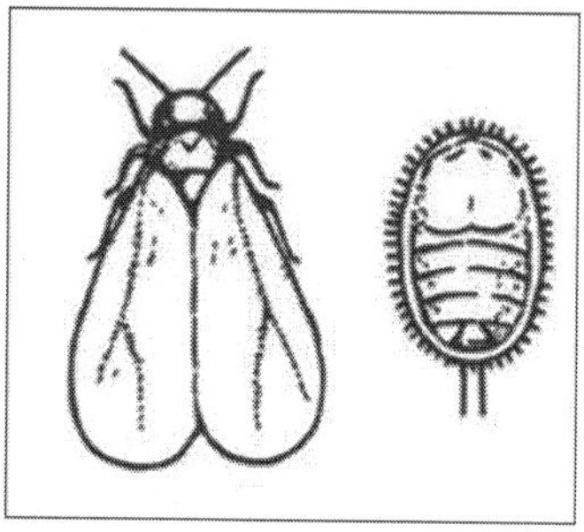

Fig. Whiteflies.

Whitefly adults resemble small gnats. They range in size from 1/10 to 1/16th inch and have four broad, delicate, milk-white wings. The immature whiteflies are attached to the underside of leaves and resemble scale insects. They are oval, flattened and yellow to almost transparent. Whiteflies often occur in tremendous numbers and when a heavily infested plant is disturbed, the air is filled instantly with a white cloud of these insects.

Whitefly: *adult (left); immature form (right)*

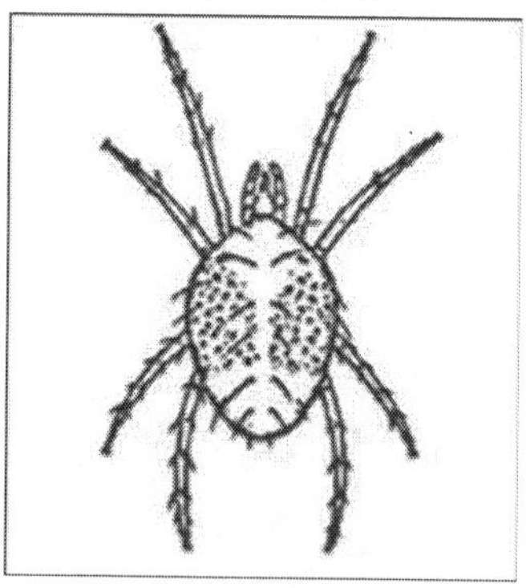

Fig. Spider Mites.

Spider mites, often called "red spiders," are most often found on the backs of leaves. They are so small they can barely be seen with the unaided eye. The adults are oval-shaped and have eight legs and no antennae or wings.

Select Recommended Plants

It is wise to choose proven performers—plants known to do well in the area intended for planting. Those known to have a history of pest problems should be avoided. Resistant plant species and cultivars should be used when available. Clavey's dwarf honeysuckle, for example, is resistant to an aphid that seriously damages other honeysuckles; and a scab resistant crab apple cultivar is a wise choice over a susceptible species. When a pest becomes a recurring problem, consider replacing the plant with a non- susceptible cultivar.

Design with Diversity

Increasing plant diversity makes it more difficult for pests to spread between plants, so it is recommended that a wide variety of plants be included in a landscape.

For example, there are four pests—the mimosa webworm, the honey locust spider mite, the locust plant bug, and the leafhopper—that can cause serious damage to honey locust stands; merely planting other tree species among the honey locusts will reduce the damage potential.

Inspect Plant Materials at the Source

Plants should be inspected for pests before they are purchased, no matter where they are purchased. New (infested) plants can introduce pest problems to the landscape.

Provide a Properly Prepared Site

Site selection is critical: The site *must* be compatible with the plants' requirements. Things to consider are exposure to sunlight, drainage, soil pH, and nutrition. The soil must be prepared carefully, using soil amendments as required for healthy growth. Lime should not be added unless a reliable soil test has indicated the need.

Plant at the Proper Depth

Planting at the proper depth may prevent damage that could make plants more attractive to pests. The hole for planting should be dug just deep enough to accommodate the root ball of the plant. Loosening soil under the root ball will result in settling and subsequent injury to the roots.

Provide Proper Fertilization and Irrigation

Plants need proper and adequate nourishment for healthy growth, and deciding what fertilizer to use and how much to apply is crucial to successful landscaping. Soil tests are helpful in determining the needs of the soil itself, but it is equally important to address the nutritional needs of each plant, as well. Transplanted trees, shrubs, and other landscape plants may benefit from a small application of fertilizer at planting, but care should be taken not to overdo it. Established plants usually receive adequate nutrients when the lawn is fertilized.

Apply Mulch Around Landscape Plants

Mulch will conserve moisture and protect plant roots from extreme temperatures; and it can prevent bark injury from lawn mowers and string trimmers.

It should be applied 2-4 inches deep and kept at least 2 inches away from tree trunks and the bases of shrubs.

Know When to Prune

There are good and bad times to prune. Caution should be taken to ensure that the pruning process does not leave trees and shrubs more vulnerable to damage from other sources.

For example, oak wilt, a serious disease of red and white oaks, is spread by a fungus which beetles carry from infected trees to fresh wounds on others; therefore, oaks should not be pruned in spring and early summer when insect activity is high. Most plants can be pruned in late winter or early spring, but trees and plants that flower in the spring should be pruned *after* the blooms fade.

Manage Non-infectious Problems Promptly

Problems transmitted from one plant to another by living organisms are called infectious; those that result from factors other than living organisms are called Non-infectious. Many landscape plant problems are caused by Non-infectious factors. < p> Plant problems resulting from soil compaction and mechanical damage to the bark often are caused by people, not pests. Environmental factors such as extreme cold or hot temperatures, drought, flooding, and lightning also can cause problems, many of which can be diminished by implementing proper cultural practices. Watering during periods of drought, improving soil drainage, and fertilizing appropriately can minimize injury from Non-infectious factors. Pesticides cannot prevent damage caused by Non-infectious factors.

Inspect for and Monitor Insects and Diseases

There are two things to remember relative to monitoring landscape plants for pests: Regular inspections are helpful in preventing serious pest damage; and early season infestations can be more damaging than those that show up later.

Physically Remove the Pest

Physical removal and destruction is one effective way to manage many pest problems. The eastern tent caterpillar is an example. This insect problem starts in a small group of eggs attached to a twig in a tree.

After the eggs hatch, the caterpillars feed on leaves at night. During the day they hide from birds in webbed tents built in the forks of tree limbs. Much of their damage can be avoided by removing and destroying these tents when the caterpillars are still small and inside. Pruning out pest-infested branches can be an effective way to avoid further damage on vigorously growing plants. Oystershell scale on red twig dogwood can be controlled in this manner.

Identify the Location of Injury

When pest damage is identified, it is important to determine which part of the plant is being affected. This will help in deciding whether or not a pesticide is needed; pest damage to plant leaves usually is less serious than damage to trunks, stems, or roots, so pesticides might be ruled out if only the leaves are involved.

Use Biological Management Methods

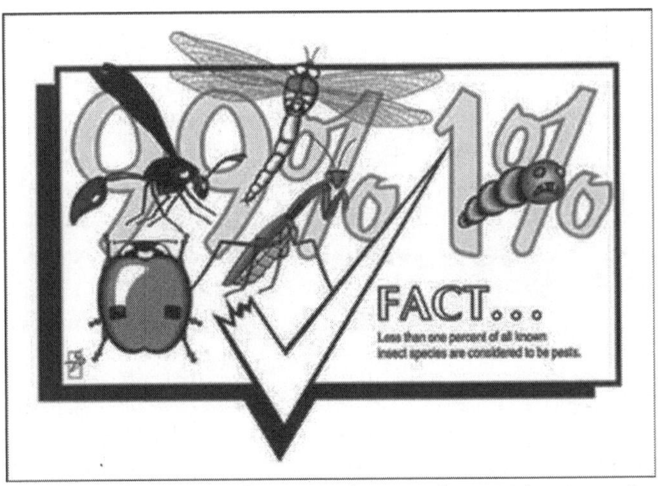

Biological control involves the use of natural enemies to reduce or prevent pest damage. Not all insects, mites, and diseases are harmful to plants; many, in fact, are predators, parasites, or pathogens. It is important to distinguish between pests and their natural enemies. When the natural enemies are present, they should be given time to become established and control the pest.

Predators attack, kill, and eat multiple numbers of pests. Parasites lay an egg in or on a pest; and when the egg hatches, the new parasite consumes and usually kills the pest as it matures.

Pathogens are free-living microscopic organisms (bacteria, fungi, viruses, etc.) that invade the pest and cause a disease that weakens or kills it.

Use Pesticides as a Last Resort

Pesticides should be used only when the pest threatens to cause serious damage to the host. When more than one pesticide is available to control the pest on the specified site, the one that is least harmful to natural enemies should be selected.

A STRATEGY FOR PEST CONTROL IN HOME GARDENS

Home vegetable and fruit gardening is a popular outdoor recreational activity in the United States. The gardener gains satisfaction in growing food crops at home. However, insects, weeds, and diseases can become a problem. The best pest management plan for home gardens is prevention. The following pest prevention methods are recommended.

Plant Disease-Resistant Cultivars

Controlling infectious diseases is difficult; therefore, emphasis should be placed on prevention. The first step is to determine what diseases are most common on the crop to be grown, where it will be grown; then, if available, choose cultivars known to be resistant to those diseases. But if disease-resistant cultivars are unavailable (or not chosen), preventative pesticide applications should be considered as a precautionary measure.

If a fungal disease is identified, a fungicide registered for controlling it on the host plant must be applied on a regular basis (usually every 7-14 days). Very thorough coverage of the leaf and stem surfaces of the entire planting is necessary since most fungicides available to homeowners are not curative, but preventative; *i.e.*, they must be applied to healthy leaves to *prevent* invasion by the fungus.

Remove Plant Residues

Since many pests can overwinter in dead plant debris, it is important to remove vegetation as soon as its produce is harvested, thereby eliminating possible Harbourage. Removing expended plant material will render the garden less attractive to insects overall, thus lessening the potential for infestation of plants not yet harvested. Weeds should be removed before they have a chance to produce seed; if the plant tissue is healthy, it may be composted. In orchard situations, fallen fruit and leaves should be collected and removed to ensure that they do not host pest populations destined to become a problem the following year.

Purchase Healthy Plants

Plants should be inspected carefully, before purchase, to be sure they are free of insects and disease. Discoloration and stunting are two signs of pest damage.

Improve the Soil

Organic matter such as compost should be mixed into the soil at the beginning of each gardening season unless the soil is heavy clay that holds too much water for long periods of time. Garden soil should be tested at least every three years and amended by fertilizing as required. A loose, fertile soil promotes healthy plant growth, and healthy plants can out-compete pests, yielding more high quality produce.

Use Proper Gardening Practices

Optimum growing conditions yield plants with fewer pest problems. Planting, spacing, watering, fertilizing, and controlling weeds according to prescribed guidelines for the plants grown in the area grown will result in gardening success.

Mulch to Prevent Weeds

A thin layer of grass clippings (not recently treated with herbicides), leaves, straw, sawdust, wood chips, or compost can be spread around plants to control weeds. As these organic materials decay, they add nutrients to the soil. Mulch helps retain soil moisture and moderates soil temperatures in the summer.

Black plastic can be used as a ground cover to conserve moisture and control weeds. It will warm the soil in the spring but may retain too much heat for some plants in summer. Plastic mulches should be used only with plantings of annual vegetables and flowers; in landscape beds, it can prevent water and air from reaching the roots of trees and shrubs. It also may encourage shallow rooting, leaving the plants more susceptible to cold injury.

Rotate Crops and Diversify Plantings

Vegetable crops should be rotated to different locations in the garden each year in an effort to reduce the potential for buildup of pest problems in the soil. But if a serious problem does occur, the crop involved should not be replanted in the questionable area for at least five years.

Encourage Beneficial Insects

Less than one percent of all known insect species are considered pests. It is important to know the difference between beneficial and harmful insects. Beneficial insects include ladybugs, bees, green lacewings, praying mantises, dragonflies, and wasps. Spiders are examples of beneficial arthropods.

Scout for Pests and Symptoms of Pest Damage

Plants should be inspected for pests several times each week. Insects feeding on plants present the gardener with two choices: Tolerate the damage, or try to control it. Control methods that don't require pesticides include hand removal of the insects, early harvest, or tolerance of small amounts of damage. For example, corn earworm damage to the tips of sweet corn ears can be cut off; the outer leaves of cabbage damaged by insects can be removed; superficial blemishes on the skin of an apple can be peeled off.

An insecticide application may be necessary when plants are in danger of being severely damaged or destroyed. But before using a pesticide the home gardener should understand that

- The pesticide label is a legal document;
- It is the user's responsibility to read and follow the label explicitly;
- Pesticides must be applied only at rates specified on the label (more is not better);
- Pesticide labels may stipulate waiting periods—called preharvest intervals—which must elapse between application and harvest of the crop for human consumption; fruits and vegetables always should be washed thoroughly before eating.

MANAGING DISEASES AND PESTS IN HOME VEGETABLE GARDENS

Vegetable gardening is a rewarding and enjoyable hobby that provides a supply of fresh produce in season and that also can help to reduce a family's food costs. Vegetable crops are threatened by various disease, insect and weed problems that may reduce the yield and quality of the produce and even destroy the plants. Diseases and pests can damage vegetables from the time seeds are planted until after the crops are harvested, and gardeners who fail to follow good growing practices that minimize damage from diseases and pests in their gardens will suffer many disappointments.

The impact of plant diseases on vegetables often appears less direct than that of insects and weeds. Insects usually can be seen with the naked eye, and their activities quickly noted, while weeds are conspicuous, competing directly with vegetable plants for space, nutrients and moisture. Diseases, on the other hand, may make their presence known only when plant growth begins to slow down or when productivity is less than expected. No causal agent may be visible, only symptoms indicating that something is wrong.

Based on general cause, there are two kinds of diseases: biotic and abiotic. Biotic diseases are caused by an identifiable microorganism or infectious agent, for example, bacteria, fungi and viruses. Abiotic diseases result from unfavourable growing conditions or environmental stresses, such as extreme temperatures, too much or too little water, and nutrient imbalances. Both kinds of diseases exist widely in vegetable gardens across Canada.

Plant-feeding insects, mites and nematodes become pests if they injure crop plants sufficiently to interfere with the capacity of the plant to produce food. Some species of insects also transmit plant pathogens. Aphids, for example, transmit several viruses, and cucumber beetles can spread bacterial wilt and squash mosaic virus. The presence of some plant-feeding pests can be tolerated, but control may be necessary if the pests begin to cause significant damage. Home gardeners usually tolerate more pest injury, especially cosmetic injury, to their vegetables than do commercial producers. Many pest species are kept at a low level by natural enemies, such as predators and parasites, and additional control measures may not be required. It is important to realize that the population build-up of natural parasites and predators lags behind that of the pests, so some damage usually is evident. Special practices may have to be employed to keep the pest population at a level where damage will not be serious.

The most effective pest control strategy for the home gardener utilizes a combination of practices; these include planting resistant cultivars, providing optimum soil conditions, fertility and moisture supply, selecting pest-free seed and transplants, employing crop rotations, monitoring carefully for pests, destroying plant residue, cultivating, mulching, and using pesticides only when necessary. Such practices often keep the populations of garden pests at tolerable levels. The role of these and other techniques in helping to minimize disease, insect and weed problems in home vegetable gardens.

MONITORING

Traps and lures: Baits and lures using light, colour, sex attractants (pheromones) or food to draw a pest to a trap may be effective for some pests.

Wireworms can be trapped by burying pieces of potato or carrot in the soil, then checking the bait every few days for the insect. Infested baits can be collected and destroyed. Slugs can be drawn to reservoir traps of beer or to a mixture of molasses, water and yeast. Boards laid out in the garden can be used to trap slugs because they look for a cool, damp spot to hide during the day. These pests can then be collected and destroyed daily. Earwigs can be trapped in wooden traps or in containers filled with water containing a small amount of detergent. Growing an insect pest's favourite plant as a trap crop to lure it from the garden also works. However, the pests must be controlled on the trap crop or they will migrate back to the garden.

Cultural Practices

Healthy seeds and transplants: Many disease-causing organisms are capable of living in and on vegetable seeds; therefore, it generally is unwise for a person to save seed produced in the home garden. Rather, it should be purchased from dealers who have a reputation for producing or selling high quality, disease-free seed. When growing transplants, gardeners should use high quality seed and a pasteurized growth medium. If transplants or rootstocks are purchased from a supplier, they should be carefully checked for signs of pests and diseases, and any found to be infested should be destroyed.

Crop rotation: Moving the preferred host plant of an insect or disease organism hampers that pest's ability to feed and reproduce. Many disease-causing organisms do not survive long in the soil in which a different crop is planted.

Exceptions include fungi that cause such diseases as fusarium wilt on cabbage, potato or tomato, and clubroot of cruciferous crops; once the soil is infested with these pathogens, it can remain so for several years. Many insects and mites hibernate or lay their overwintering eggs in the soil near their preferred hosts. Moving the garden to a different location, or even switching the host crops from one to another part of the garden, has the effect that, upon emergence in the spring, the pest may find a non-host plant on which it is unable to feed and reproduce. This practice also may help to control weeds, especially if part of the garden is summer fallowed for at least one growing season. Crop rotation has the added advantage of allowing the soil to rejuvenate itself. Cabbage, turnip and potato use large amounts of nitrogen and should be preceded with legumes, for example, pea or bean, followed by fallow and the incorporation of compost or manure.

It is advisable to rotate among vegetable families because different members of a plant family often are susceptible to many of the same pest problems. For example, tomato, potato, eggplant and pepper are in the potato family and share many common diseases and insect pests, as do members of the cucurbit family (cucumber, melon and squash) and the crucifer family (cabbage, broccoli, Brussels sprouts, cauliflower, kohlrabi, turnip and radish). It is desirable to leave as much time as possible between related crops; three to six years is ideal. If rotation is not practical, gardeners should at least alternate the cultivars of vegetables that they are growing, choosing pest-resistant ones whenever possible.

Manual and mechanical methods: Hand-picking large, slow-moving pests, such as caterpillars, Colorado potato beetles and slugs, and dropping them into a container of soapy water or a 5 per cent isopropyl (rubbing) alcohol solution is effective. Shaking the plants or hosing them off with a spray of water dislodges some insects and mites. Barriers are meant to keep a pest away from the crop that it may harm.

Some examples are floating row covers, copper strips, abrasive materials and mulches. Used cans or frozenjuice containers opened at both ends and placed around transplants will protect them from subterranean cutworms. Copper strips repel slugs because their slimy coating interacts chemically with the copper.

Applying bands of abrasive materials, such as diatomaceous earth, stone dust or crushed egg shells, around the perimeter or between rows in a garden helps to deter slugs and crawling insects. Diatomaceous earth is an insecticide made up of the ground shells of tiny sea creatures called diatoms. The silica daggers pierce the skin of the insect, causing dehydration. Soap solutions smother the pests to which they are applied. Tar paper collars placed on the soil around individual crucifer seedlings will prevent root maggot flies from depositing their eggs at the base of the plants.

Mulches provide an effective means of weed control in vegetable gardens. Organic materials, such as weed-free straw or grass clippings 7 to 10 cm thick, will inhibit germination of weed seeds brought to the surface by cultivation, retard development of weed seedlings, conserve moisture, and help to maintain a uniform soil temperature. After the crop is removed, the mulch can be incorporated into the soil to enhance its organic matter content. Plastic mulches increase the soil temperature and also conserve moisture. Black plastic mulches, with openings for the crop, will prevent weed growth, except at the opening.

Clear and white plastic mulches, however, permit weed growth, as some light can penetrate through the plastic. Plastic mulches are best suited for warm-season vegetables. Mulches also may prevent soil from being splashed onto plants, thereby keeping the produce clean and reducing the risk of spreading some diseases. Weed mats made of woven fabric allow water and air to reach the soil but screen out light and act as a barrier to emerging weeds, thereby providing effective, long-term control.

Removal of infested plant material: Many fungal diseases can build up and spread rapidly through the production of millions of spores. Prompt removal or destruction of diseased plan residues retards the spread of disease-causing organisms. This is an effective practice with such diseases as gray mold, powdery mildew, and various leaf spots and fruit rots. Removing diseased plant material from the garden at the end of the season eliminates an important source of inoculum for the following spring. The remaining residues should be rototilled or spaded into the soil to destroy disease organisms and to expose overwintering pests to the elements and to predators. Composting diseased plant material is not recommended because it may not destroy all of the disease organisms, even if the compost heats properly and is turned frequently. In some cases, leaving the remains of insect-infested plants in the garden may help to increase the population of beneficial parasites.

Good growing practices: When watering vegetable gardens, allow the soil surface to dry before another watering. A void frequent, light waterings as they tend to promote disease development and favour the germination of weed seeds. Watering with soaker hoses or in-ground furrows may reduce disease incidence because the foliage stays dry. When using overhead sprinklers, gardeners should be sure that the water is applied during the late morning or early afternoon.

This allows for rapid drying of the foliage. Prolonged leaf wetness favours the development of most foliar diseases. Wide row-spacing ensures rapid leaf drying and reduces the spread of certain pests and diseases by direct contact. However, wide spacing also reduces shading of the soil, thus enhancing the loss of soil moisture and favouring the growth of weeds. Keeping plants as healthy as possible is an important strategy in managing insect and disease problems. Planting into a warm, moist, well-prepared, well-drained seedbed, maintaining a high organic matter content, applying fertilizers correctly, and not cultivating when either the plants or the soil is wet, are recommended practices. Monitoring the crop through the growing season may help to keep pest problems small by facilitating prompt control. Accurate record keeping is suggested to ensure success from year to year.

Companion planting: Many insects prefer to feed on plants belonging to specific families and reject others. For example, the imported cabbageworm feeds on cabbage, radish, kohlrabi and other cole crops; therefore, interplanting unrelated plants, such as onion, with cole crops will help to deter feeding and limit damage by this pest. Furthermore, garlic, onion and other aromatic plants interplanted among other garden crops are believed to repel certain insect pests. Alternating vegetable cultivars or species also reduces the chances of disease spread. Where root-knot and root-lesion nematodes are a problem, interplanting with French marigold or African marigold may help to reduce the populations of these pests.

Altered planting times: Seeding dates of vegetables can be altered to allow the plants to grow at times when they are less likely to be injured by certain pests. This requires knowing the life cycles of the common pest species in the area. In some cases, it may be feasible to plant certain vegetables before or after the pests have passed through their most active feeding period. For example, in Ontario, onion maggot eggs are laid near onion plants in May, and maggots soon attack plants and may cause them to die. However, onion sets planted after June 1 will escape most first-generation maggots.

Resistant cultivars: Some vegetable cultivars are resistant or tolerant to certain pests and diseases, and growing resistant plants is an excellent way to decrease the risk of damage. Certain disease-causing fungi can survive for many years in the soil and may not be managed by routine cultural practices; therefore, planting resistant cultivars is the best means of controlling them. Packages of tomato seed marked VFN produce plants that are resistant to the *Verticillium* and *Fusarium* fungi that cause wilt diseases, and to root-knot nematodes that may attack and cause galls on the roots. Other diseases best controlled by planting resistant cultivars include cabbage yellows, potato scab and cucumber mosaic. Certain vegetable cultivars also are less vulnerable to insect damage than others. For example, Red Pontiac potato is more resistant than other potato cultivars to tuber tlea beetle, and Champion radish seems to be somewhat resistant to the crucifer flea beetle.

Biological control: Beneficial insects can be attracted to the garden by planting some of their favourite nectar- and pollen-producing plants in the garden or nearby. Members of the carrot (dill, caraway, fennel and parsley), mint (catnip, hyssop and lemon balm), and daisy (yarrow and coneflowers) families are known to attract such insects. Providing a source of water and shelter also helps to keep beneficial insects around. Birds and toads also aid in the control of insect pests.

Care must be taken when applying pesticides because beneficial insects can be killed as well as the target pest. The bacterium *Bacillus thuringiensis,* or Bt, is a pathogen of certain insects and must be ingested by the larva of the insect to be effective. Only butterfly and moth larvae are susceptible to this product, although a strain of Bt has recently been developed that will kill the Colorado potato beetle. Various formulations of this microbial pestcontrol product are available for controlling larvae of certain moths and buttertlies on home garden vegetables. This and other treatments must be applied when the pest is most susceptible.

Chemical control: *Seed treatment* Planting fungicideor hot-water-treated seed helps to ensure good stands and may avoid the need to replant. Seed treatments kill diseasecausing organisms on the seed and help to protect the vulnerable seed and young seedlings from certain soil-borne disease organisms.

Treated seed is available to the home gardener and is marked 'treated' on the package. Fungicide-treated seed is usually red or some other easily identifiable colour. If untreated seed is used, it should be certified disease-free or be hot-water treated. Seed can be treated by the gardener, using recommended chemicals according to the manufacturer's directions. Small packets of seed can be treated by tearing off one corner and putting about twice as much chemical in the packet as can be picked up on the first centimetre of the flat end of a toothpick. The packet should be shaken until the seed is thinly coated. *Foliar treatment* Most foliar diseases can be controlled by spraying or dusting plants with an effective fungicide as a preventative treatment. Protectant fungicides work on the plant surface to protect against infection, but they cannot cure established infections.

If a considerable amount of disease is present, it is usually too late to apply a fungicide treatment, except to protect newly emerging leaves. Fungicides should be applied at 7- to 10-day intervals, or as directed by the manufacturer, with reapplication after rain or watering has washed the material away. A thorough covering of the plants is necessary for proper disease prevention. Early detection, timely fungicide applications and removal of diseased leaves are necessary for effective control of many foliar fungal diseases. Gardeners can choose from a wide variety of organic and inorganic fungicides for use in combatting vegetable diseases in home gardens.

The home gardener also has a number of insecticides and miticides to choose from. Most botanical insecticides are derived from plant parts and will break down quickly into harmless substances. Botanical insecticides include pyrethrin and rotenone. Insecticidal soaps are effective against aphids, but they also are non-selective and may upset the natural balance of beneficial insects and their prey. Chemical control of weeds usually is not practical for small home gardens but can be employed for larger gardens and for market gardens. If the decision is made to use a herbicide, extreme care should be taken to avoid accidental drift onto non-target species. In addition, residual soil herbicides should not be used where residues may carryover and affect succeeding crops.

Growers who apply pesticides in their home gardens should use only products that are registered and recommended for the crops being grown and for the specific diseases or pests that are prevalent in the area. It is very important to carefully follow the manufacturer's directions to ensure maximum effectiveness of a pesticide and to minimize potential adverse effects, such as poor control, crop injury and unacceptable chemical residues on the edible produce. Spot treatment is preferable to a general application because it lessens the risk of harming beneficial organisms, humans and pets.

7

Genetic Resources in Horticulture

GENETIC DIVERSITY, BREEDING AND UTILIZATION OF CITRUS FRUITS

Citrus fruits are the third largest of the fruits grown in India with an estimated production of 2979.11 thousand m.t. from an area of 369.56 thousand ha. The country is the home of many citrus species, namely, Citrus inchangensis, C. latipes, C. macroptera, C. assamensis, C. aurantium, C. jambhiri, Citrus limonia, C. karna, C. pennivesiculata and C. maderaspatana. The diverse ecogeographical distribution and the occurrence of spontaneous mutations and natural hybridization have given rise to a wide range of variability in citrus and related genera. Genetic resources of citrus in India have been well reviewed.

In India, more than 500 citrus varieties are available of which, mandarin, sweet orange and acid lime are commercially grown with regional preference considering the agroecological conditions, resulting in their increased area and production. Increase in production over the base year of 1961 is over 242.59 per cent with an annual growth rate of 7.82 per cent. Citrus fruits thus occupy 11.53 per cent of the total area under fruits in India. However, genetic base of cultivars is very narrow and enrichment of germplasm needs greater attention.

DISTRIBUTION OF DIVERSITY

There are three major centres of diversity. The types found include the papeda, pummelo and their hybrids, Citrus indica and many types of citron, lemons and mandarins. These include interesting types like Nemerutenga, Soh-Systeng, a sour fruit similar to the sweet lime and Soh-Siem, a sour mandarin. Diversity of this region is well studied and described by Bhattacharya and Dutta. In South India, the indigenous types include Gajanimma or Baduvapuli, Kichli and some wild mandarin types, *viz.*, Kodakithuli, Billikichili, Nakoor lemon, Mole puli (sour orange type), etc. In western India, at the foot of the Himalaya, the hill lemon (galgal) and Attani are common. Citron, Citron lemon, Karna Khatta, Rough lemon, Rangpur lime, acid lime, hybrid pummelo and various types of mandarins are found all over the country.

Wild types or plants growing without any attention are largely found in the foot hill regions of northwestern India, in the northeast, in south in Malabar hills and in the Western Ghats under varying ecological conditions.

Commercial Cultivars

Although citrus fruits have been grown in India since long, their commercial production is only a few decades old. These grow in area where annual rainfall ranges from 2000-3000mm as well in the regions with precipitation of 500 to 800 mm.

They are commercially grown in Andhra Pradesh, Arunachal Pradesh, Assam, Bihar, Gujarat, Karnataka, Madhya Pradesh, Maharashtra, Meghalaya, Orissa, Punjab, Rajasthan, Sikkim, Tripura and West Bengal. Among the commercially known citrus fruits, mandarin tops with respect to area and production followed by sweet orange, acid lime, lemons, grapefruit, pummelo and others. The citrus species and varieties tried as rootstock.

Germplasm Collection

Exotic collection of citrus germplasm was started in 1940. One of the collections, Kinnow mandarin is now a leading cultivar in northwest India which has replaced local cultivars. Besides Kinnow, the other exotic collections were Valencia Late, Washington Navel, Jaffa, Malta Blood Red, Pineapple, Shamouti, Ruby Orange, Satsuma, Dancy tangerine, Clementine, Cleopatra, Wilking, Temple, Duncan, Marsh Seedless, Lisbon lemon, Sporalime, Trifoliate orange, Inchang lemon, C. hystrix and Microcitrus australis. Large number of exotic collections were established at IARI, Pusa; Fruit Research Station, Abohar; Citrus Experiment Station, Gonikoppal/Chethalli; and Citrus Experiment Stations at Srirampur and Tirupati during the fifties and sixties. Recently, large number of exotic materials have been introduced from USA, Italy, Spain and Japan at NRCC, Nagpur and HRS, Yercaud, especially of mandarins, sweet orange and rootstocks. Many exotic cultivars, *viz.*, Valencia, Jaffa and Hamlin which proved commercially successful have been recommended in specific regions. Cleopatra mandarin and Troyer citrange have been found to be promising as rootstocks for mandarin and sweet orange respectively. Tahiti lime has became commercially important in Tamil Nadu and many parts of the country.

Explorations, besides saving the endangered species, help in collecting of material including wild species and the locally developed primitive and improved cultivars for specific research needs. Northeastern region is a major region of diversity of citrus where collections were attempted during the later part of 19th century and with the turn of 20th century, surveys were carried out for collecting of herbarium specimens and identification. First systematic attempt to explore and collect the samples from the region and conserve them, and to characterize them were made by Bhattacharya and Dutta.

All the areas of Assam province were explored in about ten years, valuable species were identified, specimens of both mature fruits and inflorescences with leaves and, wherever available, both seedling and clonal progenies of each specimen were collected and maintained. Surveys were later undertaken by Singh and Singh in different parts of the country and some unknown/new species were identified. Tanaka also explored and described Citrus species of India. Singh explored the northeastern region to locate site for in situ gene sanctuary of citrus especially for the endangered species.

Chakrawar *et al.* identified two promising clones of acid lime, Vikram and Pramalini, in Maharashtra. Seedless form of Santra (Mudkhed Seedless) was selected from the variability of the local cultivar. Several forms of Musambi have been identified from Central Maharashtra. Singh *et al.* collected a thornless and seedless acid lime from Central Maharashtra. At Nagpur, seedless santra has been selected which has commercial potentiality. At Periyakulam, PKM-1 acid lime was selected which gives high yield of quality fruits. Short juvenile Poncirus was also spotted during a survey. At Rahuri, 150 accessions of acid lime were collected with desirable traits of yield, resistance to canker and leaf miner and higher summer yield. Five of these accessions have potential for commercial exploitation. One of the accessions has been selected and released as Sai Sarbati for its cultivation in Maharashtra region. Similarly, exploration of citrus orchards have also resulted in selection of superior clones of acid lime and sweet oranges at Tirupati and Rahuri.

During 1988, as a result of systematic exploration by NBPGR, in the northeastern region, C. indica and many endangered species were collected for conservation. From another mission, 94 accessions were collected from all over Orissa and these were characterized and grouped into 12 known Citrus species. However, it is apparent that efforts were concentrated on selection of commercial types and many of the species, varieties which may prove to be the donor source of important gene and not a cultivar of commercial significance, are not getting due attention. Resultantly, there is a likely chance for the loss of valuable germplasm. It is of concern that many of the valuable accessions collected in the past have been lost due to heavy disease pressure. Therefore, it is essential to give attention for the collection of diversity which is available in northeastern region, central India, in the Western Ghats and the western regions.

Characterization

Characterization of citrus genetic resources was first attempted by Bonavia who described the citrus germplasm available in India and Lushington described the Citrus species of northeastern region. Other workers have also described the Citrus germplasm. These studies were based on taxonomic observation of herbarium specimen. Systematic characterization and documentation of citrus in northeastern India was done by Bhattacharya and Dutta following the descriptors of Swingle and studying minutest details of the samples. Hodgson *et al.*, Singh, Singh and Singh, and Singh and Nath also described Citrus species. However, these morphological descriptions could not remove the confusion in classification but on the contrary, the divergent views complicated the Citrus taxonomy. Further, the chemotaxonomic approaches have been used to confirm the characterization. With use of pictorial scatter diagram for characterization, Singh and Singh used numerical classification using 145 characters. For assessing diversity, electrophoresis of proteins and enzymes has proved to be an authentic tool in characterization of citrus germplasm at gene level. Biochemical markers may help in eliminating duplicates in genepool resulting from variations in phenotypic expression due to environmental conditions.

Evaluation

Citrus genepool has been evaluated for agri-horticultural characteristics. The evaluation at different locations has helped to identify cultivars/rootstocks for different agroclimatic situations as well as identification of suitable rootstocks. Variability in mandarin and acid lime was studied to understand the genotype × environment interaction.

Several citrus species types have been evaluated for their suitabilty as rootstocks, cultivars resistant to insect pest and disease and for their processing qualities. So, by effective evaluation, it has been possible to identify and domesticate the exotic and indigenous collection for commercial adoption. Kinnow mandarin growing commercially in Punjab is one of the best example of successful introduction.

Documentation

Although documentation of genetic resources of Citrus has been done for better accessibility of information, database with central monitoring system is still lacking.

Utilization

There are ample evidences to show that genetic resources of citrus have been well utilized. In fact, the germplasm collected from exotic or indigenous sources have been the basis of commercial cultivation in different citrus growing regions in the country. By and large, the commercial cultivars are the result of selection from indigenous diversity. Owing to inherent problems of breeding, only limited attempts have been made for the transfer of desired genes through hybridization. However, natural variability existing in commercial cultivars or related species have been utilized. Extensive survey of acid lime orchards in Maharashtra state resulted in selection of two commercial cultivars *viz.* Pramalini and Vikram which are superior to existing acid lime cultivars. Ten clones of acid lime were assembled and evaluated at Periyakulam for 15 years, which resulted in selection of superior types and release for commercial cultivation. At Rahuri, 150 accessions collected from different lime growing regions were evaluated which has enabled to identify 5 superior types with respect to yield, fruit quality, bearing period and disease resistance. One of the accessions, Selection-49 has been released for growing in Maharashtra as Sai Sarbati.

Exotic germplasm *viz.* Kinnow, Seedless lime, Jaffa, Valencia Late, Marsh and Grape fruit have become important commercial cultivars in different citrus growing regions. Unshiu introduced from Japan has also promise with its short duration and dwarf canopy. Indigenous collection, Nepali Oblong lemon has proved to be a best donor for imparting resistance to canker in hybridization programme. The efforts have resulted in development of acid lime having resistance to citrus canker, a serious problem of acid lime.

The reviews of rootstock situation in India suggest that Rangpur lime is the most promising for mandarin and sweet orange in central and South India. Performance of Cleopatra mandarin has also been highly satisfactory. However in Punjab, Jatti Khatti (C. jambhiri) and Rangpur lime for Kinnow; Rangpur lime and Cleopatra mandarin for Blood Red, Jatti Khatti and Cleopatra mandarin for Jaffa have shown promise. A 25-year rootstock trial suggested Belladakithuli (C. maderaspatna) as the best for Coorg mandarin with respect to fruit yield but was not recommended due to bottleneck growth at graft union. Thus, this rootstock needs further testing. Feronia limonia proved to be highly dwarfing and precocious at Shrirampur and Tirupati, and is thus of interest for high density planting. Since, rootstocks currently employed have one or other defect, work to develop ideal rootstock having resistance to root rot, nematode, salinity with good agronomic characters, by citrus breeding was started at Chethalli wherein trifoliate orange was used for imparting resistance.

Trifoliate orange has been found to be a donor for the resistance to root rot and nematode. Several hybrids having tolerance to diseases and nematode also have promising growth characters. These hybrids are being tested for their suitability as rootstock for mandarin and sweet orange.

Taxonomic Status

The commonly grown citrus species belong to three genera, Citrus, Fortunella and Poncirus. All these genera are closely related and have inter-generic fertility. These genera are grouped under the subtribe Citrinae, tribe Citreae, sub-family Aurantioideae and family Rutaceae.

Divergent views on classification of Citrus have been expressed. Swingle and Reece recognised 16 species, while Tanaka proposed as many as 145 and later 159. Tanaka classified the forms of Citrus into two sub-genera, eight sections, fifteen sub-sections, nine groups, two sub-groups, two micro-groups and 159 species. Swingle's classification had recognized the sub-genera Papeda (containing six species) and Citrus (formerly Eucitrus) with ten species. Bhattacharya and Dutta published a monograph on classification based on the study of a large number of specimens of citrus fruits collected during 1938-48 from wide range of soil and climatic conditions up to 2000 m altitude in Assam. They followed the classification of Swingle in a general way, but accepted C. nobilis Lour., C. karna Raf., C. limetta (Risso) Lush., C. jambhiri Lush., C. megaloxycarpa Lush, and species of Tanaka that were not accepted by Swingle.

Their classification found C. pennivesiculata Tanaka, and C. limonia Osbeck to be invalid. They named a new species, C. assamensis Dutta and Bhattacharya, which is known as Adajamir in Assam. Validity of this species was, however, questioned by Singh and Nath as it differs from the C. pennivesiculata Tanaka essentially in the nature of the essential oil of peel. This was identified by Tanaka as C. medica var. alata.

Subsequently, large exotic introductions were studied by Hodgson *et al.* who recognized twenty three species of Tanaka. Singh compared characters used by Swingle and Tanaka, and proposed a key to the identification of Citrus compromising the two systems and recognized C. maderaspatana Tan. (kichli), C. unshiu Tan., C. deliciosa, C. pennivesiculata Hort. ex Tan., C. semiperflorens and C. limonia Osbeck as valid species besides a new species, believed to be a complex natural hybrid of acid lime (C. aurantifolia Swingle) and papeda group parentage. Being distinct with least resemblance to any other form of citrus(it was designated as Citrus nakoor D. Singh sp. novo).

Considering the complexity of classification based on morphological data which is difficult to comprehend in terms of pattern of variation arising due to natural hybridization and mutation, numerical taxonomy has been attempted. It has been found eminently suitable in assessing the similarities between different species for their classification into groups and their phylogenetic relationship. Use of pictorial scatter diagram in clarifying the position of hybrids, probably of bispecific origin, is also advocated. Polygonal method of presentation (after Hutchinson) of morphological data was found useful in characterising different citrus types. To confirm this, chemical constituents have been proposed as additional tools. By combination of modified almen test, ammonium molybdate test and ferric chloride test, it was possible to distinguish between Citrus limon and C. jambhiri, C. limettioides and C. aurantifolia; C. grandis and C. paradisi. Usefulness of flavonoid composition of leaves, bark and ripe fruit peel in identifying the taxa of doubtful origin was also emphasised. Chemotaxonomic studies were employed to study the variability in lemon, mandarin and related citrus types. Study of 145 morphological characters, affinity relationships, chemical data and breeding behaviour favoured the contention that only three species, C. medica, C. grandis and C. reticulata with a sexual reproduction system are in Eucitrus group. Prasad studied 71 cultivars belonging to 31 species for genetic diversity and genotype affinity using D2 analysis and leaf peroxidase isozyme. The genotypes exhibited wide variability for morphological characters. Numerical analysis and polygraph suggested some affinity at intra-specific level. Very narrow relationship between Rangpur lime, to very wide relationship between pummelo and sour orange revealed the broad nature of genetic divergence in citrus. Based on minimum generalized distances, the 71 genotypes fell into four distinct clusters. Leaf peroxidase isozyme banding pattern proved the genetical differences and distinct genotypes. The isozymes resolved the differences between seemingly similar types of citrus, particularly at intra-specific level. He also noted the close affinity among mandarin orange, sweet orange, acid lime and Rangpur lime inspite of enormous diversity.

Conservation

Germplasm is maintained under different agroclimatic conditions at Central Horticultural Experiment Station, Chethalli, Karnataka; Indian Institute of Horticultural Research, Bangalore, Karnataka; Regional Fruit Research Station, Abohar, Punjab; Horticultural Experiment Station, Bhatinda, Punjab; Division of Fruits and Horticultural Technology, IARI, New Delhi; Department of Horticulture, Rahuri, Maharashtra; Citrus Improvement Project, Tirupati, Andhra Pradesh; Citrus Experiment Station, Katol, Maharashtra; Horticultural Experiment Station, Periyakulam, Tamil Nadu; Citrus Experiment Station, Tinsukia, Assam and others. At these centres, germplasm is maintained as working collections.

Citrus germplasm is conserved in field genebanks at different locations. Though substantial diversity is conserved in field genebanks, it does not contain total representation of the variability available in citrus. Many of the germplasm collected by Bhattacharya and Dutta from northeastern India are currently not available in any of the collections. At the same time, many of the germplasm are safely duplicated at few centres owing to the fact that germplasm of one region behaves differently at other region.

In Vitro **Conservation**

Maintenance of citrus germplasm in field genebank requires large area and funds, and plants are also affected by insect pests and diseases. Therefore, *in vitro* conservation of the citrus germplasm may require attention. *In vitro* conservation technique is also used for revitalization of citrus germplasm affected by virus and virus-like diseases as meristem culture technique is found to eliminate many of the viruses. However, to conserve the germplasm *in vitro*, there is need for the research with respect to genetic stability and duration of storage, etc.

Proliferation in callus was observed even after several years by sub-culturing in medium supplemented with 5 per cent sucrose alongwith 2,4- D, NAA and kinetin each of 0.25 mg l-1. *In vitro* root tip cultures continued to retain the ability to form shoot buds which were maintained at 25°C for 8-10 months and regenerated to plants.

BIOTECHNOLOGY, GENETIC ENGINEERING AND THE SEED

"*Bio*technology," deals with *live* organisms. These may be plants or animals, visible to the naked eye, or microorganisms seen only through a microscope. Agricultural biotechnology—our focus here, encompasses various methods used to modify (or create new or modified) plant life. Such life forms may range from simple one-celled algae or complex multi-celled trees. As earlier noted, research scientists in this field use biological processes involving organisms or natural substances to develop or modify products. Their aim is to change the genetic code of living cells or seeds which is carried in the molecule known as DNA (deoxyribonucleic acid). This is the process we call "genetic engineering."

DNA is composed of two alternating chemicals joined together by two other base chemicals. The sequence of the latter determines the instructions that direct cells to make the various proteins that support life. By manipulating this sequence, corporate scientists can enable these cells or seeds to *perform new functions* or to *produce new substances*. This is done by decoding the order or "sequencing" of a gene's building blocks (*i.e.*, its chemical units), and then changing that order. The latter process is known as "splicing". Through this series of steps, scientists can add or remove traits from a plant that are desirable or undesirable. They can make crops resistant to drought or poisonous to pests and weeds. The process of adding or removing traits to or from plants becomes most controversial when it is done across species. That is, foreign genetic material is inserted into the cells of organisms of a different species. This is how "transgenic organisms" are formed. The product may be something as simple as a carrot containing milk proteins or as complex as a genetically manipulated animal and even human forms.

No one doubts that biotechnology's potential for good is immense — in both medicine and agriculture. Progress has been made in finding medicines that can cure heretofore fatal diseases. Methods of "gene therapy" are already in use whereby a "sick" gene is replaced by a healthy one.

But in addition to curative ends, experiments may be undertaken for eugenic purposes, *e.g.*, to determine the height, weight, hair colour and other traits of a fetus. Indeed, a popular test in India and elsewhere enables parents to discard unwanted female fetuses until the prized male fetus materializes. A patent pending in the U.S., patent office proposes to genetically change sperm cells so that the resulting altered animals can pass down the new traits to subsequent generations. Such experiments in animal eugenics will undoubtedly pave the way to human eugenics. Hitler's dream of a "superior" race may be realized within a century of his death

The industries most active in these kind of genetic engineering methods are in the biotech, pharmaceutical and chemical fields. The top companies in these industries include, in addition to Monsanto, DuPont, Bayer, Upjohn and others. Once they have identified and patented plant genes (as well as animal), they will be able to control agriculture and the increasingly powerful food processing industry which manufactures much of the food products that we put on our tables, especially in the Western world. But as these giant TNCs expand their operations and with that their influence, in the economies of Third World countries, their agricultural production and food consumption patterns will undoubtedly reflect that influence and power. For they will have acquired patent monopolies of not only the "discovered" genes but also of the genetic engineering methods used to manipulate these genes. Inherent in this kind of experimentation is the familiar dilemma of humankind of "letting the genie out of the bottle." It's difficult to put it back in. This dangerous scientific game with genes will create transgenic forms that will not only reproduce, but in the process of reproducing could be transformed, and having been transformed, could literally move on — by wind, rain, pollen and other natural ways, to other crops or to wild plants. Eventually, these novel organisms could displace the existing forms, homogenizing agricultural life.

It is safe to say that it is in the agricultural field that genetic engineering is most popular. The easily anticipated reason is the profits it promises. Corporate scientists in agribusiness stress the remarkable increase in yield that they claim will end starvation and malnutrition. But as with any new technology—from the wheel to nuclear power—there is a fine line between good and evil. It is that fine line that is stressed by critics who fear that the evil consequences may overwhelm the good. Whether they are pessimists, as some say, or realists, as viewed by others, the end results they foresee do not bode well for humanity and especially for the "wretched of the earth."

Since the bulk of research money comes from agribusiness, the research will focus on foods that sell well in the DC markets. As has already happened in many countries in the South, arable land is increasingly allocated to cash crops that can be exported and earn foreign exchange. Moreover, farmers who used to grow several products are now inclined to grow one or two crops on the same large landholdings. The promise of huge yields from genetically engineered seeds lures even the small-scale farmers who make up the bulk of the farming community in the Third World. But once hooked on such seeds, they are locked into the contract they were required to sign when they bought these seeds.

Cargill "generously" gave free hybrid seeds to farmers in the Philippines. Instead of the increased yields and lower use of chemicals promised by Cargill, quite the opposite happened. The seeds were not suited to the moist climate of the Philippine Islands, and more pesticides, weed killers and fertilizers had to be applied. Worse still, owing to the restrictive terms of the contract signed, they were prevented from using their neighbours' seeds, as traditionally done when a crop failed. Even worse was the contract sold to Indian farmers by Monsanto which demanded a hefty 'technical fee' per acre in addition to the amount paid for the seed. But they were also locked into Monsanto's chemicals as well, and were subject to a fine if they used someone else's chemicals. Activist Vandana Shiva's description of Monsanto's methods as totalitarian is clinched by the provision that allows Monsanto inspectors to visit these farmers (even in the absence of the land owner) in order to determine whether they have violated any of the terms of the contract.

Beneath a façade of official concern about the effect of invading "seed giants" on Indian agriculture, there is a lot of support on the part of Indian agribusiness and allied interests with ties to outside agricultural ventures for opening up India to foreign agribusiness. It is argued that such ties will bring the technology and capital needed to make Indian agriculture more productive than it is now. Mixed signals have come from Seattle regarding the Indian Delegation's policy stance during the doomed Conference. What bears watching is the evolution of the BJP Government's agricultural pronouncements in the aftermath of the Seattle debacle.

Thus far, these have not been reassuring. The status of a proposed law, the Protection of Plant Varieties and Farmers' Rights Bill, remains unclear. It has been reported by the Press Information Bureau of India that this Bill was introduced in the Lok Sabha on December 3rd. But the same agency announced on December 15, 1999 that this Bill was "likely to be introduced shortly in Parliament. " Nevertheless, the draft that has been circulated has raised more questions than it has answered. One critic states that this bill "will end up benefiting the large seed corporations, some very large farmers, and corporate farming agencies." Such a Bill, if enacted, may well signal the nature of relations we can anticipate between India and powerful industrial nations like the United States — the principal architect of the new corporate order. Will India submit to "the inevitable" or will it stand and fight?

KEEP OUT PATENTS ON CONVENTIONAL SEEDS AND ANIMALS

For several years, patents on genetically modified seeds and animals have been granted worldwide. The damaging impacts on farmers, who are deprived of their rights to save their seeds, and on breeders who can no longer use the patented seeds freely for further breeding, are well known.

In Canada and the U.S., for example, the multinational seed company Monsanto has sued many farmers for alleged patent infringements. The same company has also filed court cases against importers of Argentinean soy to Europe. Furthermore, the possibility of patenting seeds has fostered a highly concentrated market structure with only 10 multinational companies controlling about half of the international seed market. Many farmers organisations and NGOs around the world are fighting against these patents. Because genetically modified organisms (GMOs) are still not grown in most countries, or only used in a small number of crops, the negative impacts of these patents are not being felt everywhere. However, there is an alarming new trend for patents not only to be claimed on GMOs (such as Round-up ready soybeans), but also on conventional plants. For example, patent claims have been made for soy beans with a better oil quality covering parts of the plant genome when used in conventional breeding and technologies to improve conventional breeding (such as marker assisted breeding).

Some of the most threatening examples in this context are patent applications from Syngenta which claim huge parts of the rice genome and its use in breeding of any food crops that have similar genomic information to rice (such as maize and wheat).

The European Patent Office has also granted a patent on aphid resistant composite plants which are based on marker assisted breeding. Other recent patent applications by Monsanto on pigs are also related to normal breeding methods, indicating the increasing danger of agricultural genetic resources becoming monopolised by a few multinationals on a global scale.

Soon the Enlarged Board of Appeal of the European Patent Office will decide on another patent of this kind - for a method of increasing a specific compound in *Brassica* species.

This decision will determine the patentability of conventional seeds in Europe.

Whereas patents on conventional plant varieties are normal practice in the U.S., many other countries, especially developing countries, do not grant patents on plants or animals. But as the recent history shows, the standards defined and used at the European, Japanese and U.S., patent offices influence international regulations (the WTO agreement on trade related aspects of intellectual property rights, TRIPS, and the World Intellectual Property Organisation, WIPO). Patent offices all over the world are pushed to adapt their regulations and practices either through the international regulations or by bilateral agreements. India, for example, has just passed a third patent amendment in order to adapt its law to the TRIPS regulations.

This frightening new trend in patent policy will affect many more farmers and breeders, than has been the case with GMO patents. Any remaining farmers rights and breeders' access to plant varieties and animal breeds for breeding purposes, will disappear everywhere. These patents will destroy a system of farmers' rights and breeders' privileges that has been shown to be crucial for the survival of farmers and breeders, for food sovereignty, and for the preservation of bio-diversity in agriculture. The vast majority of farmers in developing countries are small-scale farmers, completely reliant on saving and exchanging their seeds. In order to secure the continued existence of independent farming, breeding and livestock keeping and hence the food security of future generations, we, the undersigned farmers, researchers, breeders and civil society organisations from all over the world, restate our rejection of any patents on life, and urge policy makers and patent offices to act swiftly to stop any patents being granted on conventionally bred plants and animals and on gene sequences for use with conventional breeding technique, as well as on methods for the conventional breeding of plants and animals. We also urge companies not to apply for any patents of this kind.

Savage Garden: Slack-Potting the Dewy Pine

Over twenty years ago, British nurseryman Adrian Slack originated a rather clever way to successfully grow the often temperamental dewy pine (*Drosophyllum lusitanicum*). Glaringly omitted from my own book, The Savage Garden, I wish to correct that oversight here, for this method, called "Slack-potting"—a term coined by Barry Rice—is an ingenious method of cultivating *Drosophyllum* with long-term ease.

We must remember that dewy pines are not wetland plants like most other carnivores. Instead, they are typically found on sandy, gravel slopes that are dry most of the year, and usually alkaline in pH. The plant is found in widely scattered sites in Portugal, Spain, and Morocco. The climate is warm-temperate and Mediterranean, which means that most of the rainfall occurs in the cool winter months, with very rare frosts. Summers are warm and dry, but plants near the coast can experience early morning fog drip, which may be absorbed by their leaves.

Seed is usually the only way to propagate *Drosophyllum*, and since plants despise root and stem disturbance they are never sold through the mails. Seed is usually produced in late spring or early summer and can germinate readily if very fresh. Older seed benefits from several methods of pretreatment. You can scarify the hard seed coating by rubbing the black, egg-shaped seeds with sandpaper until the whitish interior becomes slightly visible. You can also soak the seed for a day or two in a cup of water in which you have added a tiny bit of powdered gibberellic acid—about 0.5 cm (1/8 inch) collected on the tip of a toothpick works well. I have also scarified the seed and then soaked them in a solution of one-drop Superthrive™ per cup of water.

Slack warns against moving freshly germinated seedlings about, and prefers to sow a few seed directly in their permanent pots, allowing only one plant to grow by removing any others that germinate. This can be a hassle and waste of seed if you have more than just a few to sow, but Slack's warning is legitimate since the first tiny root to appear from a seed will rapidly plunge into the soil medium and can be killed if disturbed. I prefer to sow numerous seed on damp vermiculite (they will germinate on any damp substance, even a sponge) and I check them daily after a week or two. Now this is where the Slack-potting begins. Take a 10-15 cm (4-6 inch) pot made of unglazed terra-cotta clay. Place a wad of damp, long-fibreed sphagnum moss through its drainage hole, totally blocking it. As a soil medium, I use one part each perlite, vermiculite, and horticultural sand to be excellent. Slack includes a portion of John Innes Compost #2 (a humus-based house plant soil), but I find this component to be unnecessary. (It is only available in Britain.) Dampen the mix with purified water and fill the clay pot firmly with it to the brim. Gently place your germinated seed on the centre surface of this, one seedling per pot. Place the pot in a shallow water tray in a sunny place with good air circulation, keeping the soil damp to wet for several weeks as the plant establishes itself. Let the baby plant grow for several weeks, as Slack recommends, before proceeding with phase two. The plant will grow rapidly and when it has several leaves you are ready to Slack-pot it.

You will need another pot with drainage holes at least two inches larger than the pot containing the plant. Slack recommends another clay pot, but in hotter, drier climates like California, I use plastic. Block the drainage holes with long-fibreed sphagnum. Fill the pot with the perlite-vermiculite-sand medium only as deep as to allow the smaller potted plant to sit comfortably on the medium and have its rim at the same level (or just slightly above) as the larger pot. In the gap between the two pots, tightly pack long-fibreed sphagnum.

The whole reason for this elaborate double-potting is to keep the soil around the stem and upper roots on the dry side, or just barely damp, since the prostrate stems tend to rot if left on permanently wet soils. Initially you can keep the whole double pot in a shallow tray of water as well as gently watering the plant itself, keeping the soil damp. But in a very few months the roots of your dewy pine will enter the soil of the larger pot through the sphagnum-crocked hole of the smaller. You can then water the plant via the sphagnum-moat that separates the two pots. Alternatively, you can set the Slack-potted plant in a shallow water tray, and water by the tray system only enough to keep the sphagnum-moat damp.

Just enough moisture will permeate the inner clay pot to keep your dewy pine happy, and the deepest roots will find plenty of moisture in the medium of the larger container. This probably recreates the natural situation well, for in the wild dewy pines have roots that are extensive, obtaining moisture from deep underground. Some also believe the plants absorb moisture in summer from condensing early morning fog, but misting the plant is not required. Slack-potting a dewy pine works best for greenhouse cultivation, particularly in humid climates with long, cool winters, such as Slack's own England. It is not quite as necessary if you live in a Mediterranean climate similar to the dewy pine's.

I have seen photographs of huge *Drosophyllum* plants grown in giant clay pots outdoors in Australia. Geoff Wong's prize winning plant photographed in

The Savage Garden was a judiciously watered specimen grown in a plastic pot of primarily perlite and a tiny bit of peat, then placed in that lovely glazed ceramic ornamental. I myself grew one outdoors on my deck in northern California, using a plain clay pot. The plant survived months of winter rain, a hard freeze, and baking summer heat. So while not for everyone, for many growers Slack-potting is ideal. At my nursery just this past week we Slack-potted several dewy pines. In my imagination, I visualized Adrian visiting us in his white suit and Panama hat, nodding his approval as we shared a glass of wine.

Experiments with Highland *Nepenthes* Seedlings

One of the most difficult things to describe to someone is how to grow a plant that is restricted in its tolerances. Partly to blame is our human perception of these conditions. What exactly is "bright light"? What is "clean water" and how acidic is "acid"? Highland *Nepenthes* are often one of the more perplexing groups to work with. What seems to work for one person fails when someone else tries to reproduce the cultural environment. *Nepenthes villosa*, for example, is not an easy plant to grow from seed; it has a narrow range of conditions necessary to grow well. Beans are easy to grow, highland *Nepenthes* are not!

In order to get better insight into the needs of several species, I conducted five years worth of experiments aimed at quantifying their needs. Instead of subjective terms, I used tools to measure environmental conditions in all test groups. A light meter, conductivity meter, pH meter, recording thermometers and a high quality humidity meter were employed to put some numbers behind the findings. (Cheap meters are dangerous to rely on, so only meters that could be calibrated and tested were used.) This way, anyone could nearly duplicate the conditions that were successful, and avoid potential disaster by measuring variables before plants get put into a bad environment.

The following *Nepenthes* were used in this study: *N. burbidgeae*, *N. edwardsiana* (Tambuyukon type), *N. fusca*, *N. stenophylla*, *N. tentaculata* (Mt. Tambuyukon type and Mt. Kinabalu type), and *N. villosa*. All were raised from donated seed or very tiny seedlings (cotyledon spread 0.63 cm., 0.25 inch). Dead or struggling plants were considered indicators of exceeding the species tolerance to one or more variables within their environment. The bane and beauty of using seedlings is that they do not take long to indicate improper environments; they die quickly!

Several points deserve mention, before getting into the "meat" of this chapter. First, plants grow in a system, where the elements of culture play off each other. For example, higher light levels require greater nutrient levels, as plants need more nitrogen to eliminate photosynthetic (waste) byproducts. Stresses from lower relative humidity might be offset by lower light or more available moisture at the root zone. Thus, I am not suggesting one rigid method, when explaining a successful technique. My intention is to give benchmarks and warning signs. Second, this chapter was written as a distillation from five years (1996-2001) worth of records.

A full description of these experiments would fill several volumes of this newsletter with tedious detail. My intention is to present a simplified summary of what happened during these experiments. One example of this practical focus was to simplify the complex science of light. The usual grower purchases a readily available light source or uses sunlight. The intensity of the illumination striking the plants is determined by the light source, the distance from the source to the plants, and/or the amount of shading. The simplest way to get a reasonably accurate measurement of intensity is to use a light meter, specifying the unit of measure and the source (very important). For spectral analysis of artificial lights used herein, contact the manufacturer.

Finally, it is important to note that, unlike formal experimental design, there were no control groups, *per se*. In many of these species, seedling death is the norm. Other growers, who worked with many of the same seed lots, reported complete losses in difficult species like *N.edwardsiana*, and *N. villosa*. Thus, it was nearly impossible to assign a control group without first figuring out how to keep it alive! The main support for these findings is that significantly large numbers of plants were used. The results herein are preliminary and not conclusive; other factors (such as bacterial infection, seed viability, genetics of the donated seed, etc.) could have had an effect on these findings. Readers should consider this chapter a detailed summary by a professional horticulturist, rather than a classic, formal experiment. As far as I am aware, this is the first time that meters have been used to quantify the major (growing) environmental parameters, and that these measurements have been united in print.

GENETIC DIVERSITY

Diversity among individual plants and animals, species and ecosystems provides the raw material that enables human communities to adapt to change - now and in the future. Deprived of bio-diversity, the ability of humankind to meet the challenges resulting, for example, from climate change would be severely limited. The diversity found within the small number of plant and animal species which form the basis of world agriculture and food production remains a small but vital part of the earth's bio-diversity. Genetic diversity is not evenly distributed throughout the world; it is in fact concentrated in tropical and sub-tropical areas, where the majority of developing countries are located. Plant genetic resources are of an inestimable value, and will continue to be so in the future, independently of whether scientists use them by means of conventional plant breeding or modern genetic engineering. The importance of plant genetic resources as the ultimate source of food is enormous. Their loss constitutes a serious threat to world food security. The conservation of genetic resources goes far beyond the salvation of species. The objective must be to conserve sufficient diversity within each species to ensure that its genetic potential will be fully available in the future. Within the Plant Kingdom, around 350,000 species have been classified, out of which about 80,000 have been classified, out of which about 80,000 have been found to be edible. In the course of history, mankind has utilised about 7,000 of these plant species for food. Today, only 150 plant species are cultivated and of these, the so-called major crops can be contemplated in about 30 plant species which are producing about 95 per cent of the world's calories and proteins. About 75 per cent of food consumption comes from only 12 plant species and 5 animal species. Half of this food comes from only 4 plant species (rice, maize, wheat and potato) and 3 major animal species (cattle, swine and poultry).

Over millennia, men and women farmers have developed within each domesticated food species thousands of landraces (farmers varieties) and breeds adapted to local conditions and needs. This "human-made" agro-biodiversity is now seriously endangered. Today the world population amounts to about 5 billion people.

An increase of about 60 per cent is expected in the next thirty years, which will bring the world population to some 8 billion people. Today, 800 million people are chronically malnourished. On the basis of recent estimates, in the developing countries alone, food production will need to increase by about 60 per cent in the next 25-30 years in order to keep pace with the expected demographic growth.

PLANT GENETIC RESOURCES

Despite the useful progress made so far, somc major questions are still currently being debated in the FAO fora that deal with plant genetic resources for food and agriculture, *i.e.*:

- Who owns plant genetic resources?
- Whose responsibility is it to conserve them?
- Who benefits from their use?
- What form the benefits deriving from their use should take, and how should they be shared?
- How can the benefits of biotechnology be directed to the largest number of people and countries?
- How can measures be taken to prevent certain applications of biotechnology endangering the livelihoods of large numbers of people in developing countries? (for example by new products replacing the agricultural commodities that provide their livelihood)?
- How can biotechnological research be directed towards the maximum benefit to world society?
- What measures can be taken to prevent modern, intensive agriculture, with homogeneous varieties, stamping out the world's rich inheritance

of plant germplasm conserved in the heterogeneous local varieties in farmers' fields?

In June 1996, the Fourth International Technical Conference on Plant Genetic Resources, which was attended by 150 countries and 54 inter-governmental and non-governmental organisations, was held in Leipzig, Germany. The Conference was convened at the request of the FAO Commission on Plant Genetic Rthe request of the FAO Commission on Plant Genetic Resources, and was endorsed in Agenda 21 of the United Nations Conference on Environment and Development, and at the Nairobi Conference for the Adoption of the Agreed Text of the Convention on Biological Diversity.

The Conference adopted a Global Plan of Action for the Conservation and Sustainable Utilisation of Plant Genetic Resources, together with the Leipzig Declaration, and it also considered the first Report on the State of the World's Plant Genetic Resources.

The Global Plan of Action was prepared through a participatory, country-driven process, involving a wide variety of stake-holders: governments, non-governmental and industry organisations, and individual scientists. A total of 158 governments prepared Country Reports, assessing the status of their plant genetic resources, as well as their capacity to care for and utilise these resources. Twelve regional and sub-regional meetings were held, where governments considered regional problems and opportunities, and made recommendations for the Plan. The Report on the State of the World's Plant Genetic Resources is the first comprehensive world-wide assessment of the state of plant genetic resource conservation and use, identifying the urgent priorities for action which are, in turn, addressed in the Global Plan of Action. The Report and the Plan are now two strategic elements of the Plan are now two strategic elements of the FAO Global System for the Conservation and Utilisation of Plant Genetic Resources.

The Plan is intended to provide a coherent strategy for action in several fields: *in-situ* and *ex-situ* conservation, sustainable utilisation of plant genetic resources, and institutional and capacity building. Therefore, the GPA needs to be taken as a whole and considered in a comprehensive manner embracing all operational, programmatic, policy and political aspects.

The Leipzig Declaration asserts that "our primary objective must be to enhance world food security through conserving and sustainably using plant genetic resources". This Declaration commits the Governments to taking necessary steps to implement the Global Plan of Action.

On a different, albeit related, plane, the world's political leaders, meeting in Rome at the 1996 World Food Summit, have made a public commitment to end hunger. Through the Plan of Action adopted by the Summit, governments, international organisations and all sectors of civil society are encouraged to join forces in a concerted effort to ensure access at all time to the food required for a healthy active life for all the world's people. Meeting the same month, in November 1996, the Third Conference of the Parties of the Convention on Biological Diversity held in Buenos Aires focused on agricultural bio-diversity, and its res focused on agricultural bio-diversity, and its role in ensuring a transition towards sustainable agricultural practices.

All of these inter-governmental meetings recognised the importance of plant genetic resources for food security and called on countries to implement the Global Plan of Action for the Conservation and Sustainable Use of Plant Genetic Resources for Food and Agriculture.

The needs highlighted at Leipzig, Rome and Buenos Aires, and the recommendations agreed upon by governments in the Global Plan of Action for Plant Genetic Resources present many challenges to all crop scientists, in terms of developing or adapting suitable technologies and approaches for the sustainable utilisation of plant genetic resources for food and agriculture (PGRFA). The concept of sustainable utilisation is defined in the Convention on Biological Diversity as the use of genetic resources in a productive manner which does not lead to the long term loss of such resources. This provides for an approach which integrates both conservation and utilisation and which spans all aspects of the use of PGRFA, whether by scientists or farmers.

Plant genetic resources can be described as the part of bio-diversity that nurtures people and is nurtured by people. Horticultural crops, although constituting only one part of plant genetic resources for food and agriculture, comprise a very wide range of species of vital importance to ensuring age of species of vital importance to ensuring food security.

All governments have now recognised that the utilisation of plant genetic resources is the key to improving agricultural productivity and sustainability and can contribute to socio-economic development, food security and the alleviation of poverty. To achieve these desirable goals, the Global Plan of Action promotes an integrated strategy for the conservation and sustainable utilisation of PGRFA with the following features: *productivity* - greater use of plant genetic resources will be required to contribute to the productivity increases needed to meet growing populations. This will require continued access to, and exchange of, the world's plant genetic resources; *sustainability* - there is a need to ensure that such use of plant genetic resources is coupled with the conservation of plant genetic resources, both *in-situ* and *ex-situ*.

This will require, *inter alia*, approaches to crop improvement which allow the maintenance of higher levels of genetic diversity and resources in production systems, thereby contributing to reduced genetic vulnerability and less genetic erosion. Additionally, conservation programmes should be clearly linked with utilisation efforts and the sharing of benefits, in order to reinforce the sustainability of such programmes; *equity* - those responsible for conserving and developing plan those responsible for conserving and developing plant genetic resources should be able to participate fully in the benefits derived from their use. There is thus a need to develop crop improvement approaches to enable farmers in marginal areas, as well as those in high-productivity areas to benefit fully from the utilisation of plant genetic resources. In meeting these challenges of harnessing plant genetic resources for sustainable agriculture and future food security, the role of crop scientists will be crucial.

THE INTERNATIONAL MUSHROOM GERMPLASM CONSERVATION

Fungi are regarded as being the second largest group of organisms in the biosphere after the arthropods. The total number of fungal species is evaluated as 1,500,000 in the world. Only 5 per cent of theses species are described and catalogued. Therefore our knowledge on the worldwide fungal species is poor. Out of the 70 per cent described species of fungi, there are about 10,000 species of mushrooms in which about 2,000 species from more than 30 genera are regarded as prime edible mushrooms, but only a few dozen of them are cultivated. In the particular of tropical and sub-tropical areas, the fungal and mushroom genetic resources are certainly the most abundant of the world and the least known.

This network was established to strengthen international collaboration among specialised institutions, with a view to constituting a coordinated global system of mushroom germplasm collections under the aegis of FAO, and to facilitate technical communication and the exchange of strains of cultivated mushroom, and of germplasm of other mushroom species of interest for food and agriculture. The network also has the aim of establishing a more comprehensive and coordinated information mechanism, which would help to make mushroom strains, together with appropriate production technologies, available to as many interested countries and growers as possible.

THE USE OF BIOTECHNOLOGY

Through modern biotechnologies that are helping to overcome some of the limitations of conventional breeding, wild diversity can be more readily incorporated into crops and thus contribute to world agricultural development.

It is recognised that production must be intensified, productivity increased and productive natural systems must be suitably managed - all in a sustainable manner. This requires the combined application of new and old technologies, including innovative approaches to plant breeding and to farming practices. The success of such endeavours will depend on the sustainable utilisation of a broader range of species and genetic material within each species, including the wild relatives of domestic species.

Many of the recent advances in agricultural technology, including biotechnologies, have been made in industrialised countries where they have often been developed on a commercial basis. In this sense, the potential for the new biotechnologies to improve the livelihood and food security situation of the world's poor is largely under-realised.

The new biotechnologies do provide excellent tools for manipulating the genetic diversity of crop varieties and animal breeds among the already domesticated species and further increasing productivity and stabilspecies and further increasing productivity and stability. Through the development of plants and animals that are more adapted to ecological conditions, the stability of production can be increased. The genetic manipulation of other biological elements in crop production systems, such as soil bacteria associated with nitrogen fixation, or natural predators of pests (especially insects and to some degree pathogens and weeds) is also a tool that can be used for enhancing sustainability in agriculture. Ours is the first generation with the immense biotechnological power to radically change the evolutionary processes of natural and artificial selection. Ours, too, is the responsibility that goes with this power.

However it is recognised that there are also risks involved in the application of biotechnologies; these risks include misuse by humans, and environmental accidents. To address these and other issues, a draft Protocol on Bio-safety is now being negotiated by those countries which are signatories to the Convention on Biological Diversity. With the advent of genetic engineering, the private sector biotechnology industry has promoted the extension of industrial patenting regimes to living organisms - an approach popularly known as life patenting.

As a result, genes, micro-organisms and varieties of higher species (plants and animals), either discovered in nature or genetically manipulated by er discovered in nature or genetically manipulated by breeders or genetic engineers, can and have become the intellectual property of individuals or industry in countries in which "life patenting" has been legalised. Between 1981 and 1985 at least 1175 patents for human genes were granted worldwide; and of these patent applications, some 75 per cent were from the private sector.

Perhaps for reasons guided by the market, the socio-economic and food security needs of the poorer sections of society, especially in developing countries, have not yet been sufficiently targeted by biotechnology programmes, or have not been adequately considered in shaping research policy. This is an important point for decision-making in relation to the allocation of public sector funding for agricultural research, since it is estimated that there are now at least 1,400 million people in developing countries who are dependent upon resource-poor farming systems which have not benefited from technological advances in plant or animal breeding or biotechnology.

It can also be debated whether current intellectual property right systems allow appropriate incentives for directing public sector agricultural research so that the social and economic needs of these poor countries can be adequately met. This issue becomes particularly worrying when it is considered that the limitation of access for the world's poor farmers to improved genetics for the world's poor farmers to improved genetic materials could become a constraint to increased food production.

THE FAO GLOBAL SYSTEM

Within the programmes of the Food and Agriculture Organisation of the United Nations (FAO), the inter-related questions of who benefits from the use of genetic resources and who owns genetic resources stored in international seedbanks, in farmers' fields, or in nature, became the focal point of international debates during the 1980s. To confront and address these problems, the member countries of FAO agreed on the need to develop a global system which might ensure the conservation and sustainable use of plant genetic resources for food and agriculture, as well as the just and equitable sharing of the benefits and responsibilities derived from them. The development of the Global System on Plant Genetic Resources for Food and Agriculture began in 1983 with the establishment of its governing body, the Commission on Plant Genetic Resources, now the Commission on Genetic Resources for Food and Agriculture, a permanent inter-governmental forum for discussion and consensus building between countries. Governments also adopted, that same year, an international non-binding agreement, the first of its kind, called the International Undertaking on Plant Genetic Resources, which provides the formal frameant Genetic Resources, which provides the formal framework for the Global System. The Commission has since coordinated, overseen and monitored the development of a Global System for the Conservation and Utilisation of Plant Genetic Resources for Food and Agriculture.

Countries continue to negotiate and develop, through the Commission, other important elements of the System which include various international agreements, scientific regulations, technical cooperation mechanisms and global instruments which are in different stages of development. The availability of information about plant genetic resources of crop species is fundamental to being able to harness the genetic variability conserved in collections and critical for evaluating the usefulness of *in-situ* conservation initiatives to counter the risk of genetic erosion. In response to this need, the FAO World Information and Early Warning System on Plant Genetic Resources (WIEWS) was created five years ago, within the overall framework of the FAO Global System on Plant Genetic Resources. Its primary purpose is to facilitate access to information so as to allow a more coherent utilisation of available genetic resources for plant improvement work.

GLOBAL NETWORKS

The importance of conserving plant genetic resources has been stressed in several international agreements, such as the International Undertaking on

Plant Genetic Resources and the Convention on Biological Diversity and its Agenda 21. The successful conservation and sustainable utilisation of plant genetic resources for food and agriculture involves action by a wide range of people in every country: policy makers, planners, scientists, genebank curators, breeders, rural communities and farmers. Strong coordination mechanisms are required at the national level to enable all these players to participate constructively. In response to this need, through its international networks on *ex-situ* and *in-situ* conservation, FAO is pursuing efforts in order to coordinate, under this umbrella, the conservation and utilisation of er this umbrella, the conservation and utilisation of plant genetic resources of many agricultural crops and to facilitate the exchange of genetic material, information and technology.

The need for germplasm conservation is recognised for all crops and plants of agricultural importance. This, in turn, dictates the need for programmes in evaluation and documentation, and these activities must be carried out in cooperation with institutions, scientists and technicians with crop-specific competence. In this context, we consider it important to foster the establishment of crop-related networks with special competence and responsibilities related to genetic resources of specific crops.

The development of clearer policy lines related to the conservation of genetic variability of crop species, through regional and global inter-country cooperative programmes, was one of the principal aims of the Leipzig Technical Conference on Plant Genetic Resources. The Global Plan of Action on PGR is now under implementation through a partnership mechanism involving inter-country crop-related networks, as well as several national and international institutions and organisations, including non-governmental organisations. The Global Networks which have been recently launched by the FAO are thus elements of the Global Plan of Action.

It is being demonstrated that inter-country networking constitutes a useful instrument to facilitate policy itutes a useful instrument to facilitate policy development and to provide guidance and global coordination for inter-institutional research work concerning crop genetic resources. It should be underlined that the crop-related networks that operate under the aegis of the FAO have been constituted on the basis of voluntary and self-funding participation of interested countries, institutions and scientists. In facilitating the implementation of the Global Plan of Action on PGR, and recognising the important role of horticultural crops for agriculture and for food security, FAO has concentrated efforts on promoting regional, inter-regional and global crop-related networks, particularly oriented towards the horticultural sector, as instruments to facilitate scientific exchange, information sharing, technology transfer, and research collaboration. The networks also provide a forum for determining ways of sharing responsibility for the collection, conservation, evaluation, characterisation and utilisation of horticultural genetic resources.

These crop-related networks bring together different types of specialists to set collaborative research objectives, define policy priorities and strengthen activities on the conservation and utilisation of genetic resources in relation to specific groups of crops. Most of the partners of these inter-country cooperative programmes are internationally recognised scientists and members of the Internatignised scientists and members of the International Society for Horticultural Science which further strengthens collaboration between the ISHS and FAO.

The innovative aspect of these networks is that they promote a co-ordinated approach to identifying, evaluating and conserving the genetic variability of selected crop species, with the aim of improving cultivars and adapting them to farmers' needs. The networks combine a thorough knowledge of the agricultural conditions of farmers in network member countries, with an understanding of the genetic potential of the crop species in question, and use a "farmers-to-farmers" basis for the development of conservation and utilisation activities. That involves the collection from farmers of locally adapted germplasm, its improvement and its return to the same farmers, or farmers in similar biotopes. The work of several of these networks is also guided by an economic intelligence function (analysis of market factors), which helps farmers to improve quality and suit their production to market requirements.

The distinguishing characteristic of the networks' components on plant genetic resources consists in identifying the genetic variability within each particular crop species, and the development of methodologies for characterisation, evaluation and conservation of different genotypes in view of their utilisation by present and future generw of their utilisation by present and future generations of farmers. As such, the networks are regarded as very useful instruments for facilitating the process of implementing the Global Plan of Action, and their expansion, in terms of both crops and regions covered, is being pursued. The crop-related networks also provide a useful forum for updating the Report on the State of the World's Plant Genetic Resources. Computerised data bases are being developed for several of the networks and the information being gathered is available to be systematically included in the World Information and Early Warning System.

The many crop-specific networks and working groups may operate at sub-regional, regional, inter-regional or global level. The networks that FAO has developed over the last seven years, with the aim of promoting a coordinated approach to identifying, evaluating and conserving the genetic variability of selected crop species, include the Global Mushroom Germplasm Conservation Network; the Olive Genetic Variability Conservation Network; the International Network on Cactus Pear; the Mediterranean and Inter-American Citrus Networks and the Global Citrus Germplasm Network; the Inter-regional Cooperative Network on Nuts; the Global Network on Tropical and Subtropical

Fruit Genetic Resources, and the Network on Traditional Crops of Southern Africa.

CONCLUSION

The rural poor depend upon biological resources for an estimated 90 per cent of their needs. In the industrialised world, access to diverse biological resources is necessary to support a vast array of industrial products. In the continuing drive to develop efficient and sustainable agriculture for many different conditions, these resources provide raw material for plant and animal breeding, as well as for the new biotechnologies. Genetic diversity in agriculture enables crops and animals to adapt to different environments and growing conditions. The ability of a particular variety to withstand drought or inundation, grow in poor or rich soil, resist one of the many insect pests or diseases, give higher protein yields or produce a better-tasting food are traits passed on naturally by its genes. This genetic material constitutes the raw material that plant and animal breeders and bio-technologists use to produce new varieties and breeds. Without this diversity we would lose the ability to adapt to ever-changing needs and conditions. Sustainable agriculture could not then be achieved in many of the world's different food production environments.

8

Greenhouse Crops System in Horticulture

NUTRIENT AND FEEDING GUIDELINES IN HYDROPONICS

All plants need nutrients to supply them with the elements needed for vital biochemical processes. Nitrogen (N), phosphorus (P) and potassium (K) are the top three generally listed, but there are more than a dozen others. Magnesium (Mg), iron (Fe), calcium (Ca) and several more perform essential roles in the life of hydroponic plants, just as they do in soil-based gardens.

Nitrogen is used by growing leaves. But despite the fact that the air is about 79 per cent nitrogen, plants need it in the form of a supplement. The N2 molecule in air is very stable and plants don't break it apart to use single nitrogen atoms. Phosphorus is essential to root growth. Potassium aids in disease resistance owing to its role in enzyme formation.

The other elements perform a variety of functions. Calcium, for example, is a large component of cell walls and also helps deliver ions to various parts of the plant. Chlorine (Cl) is a component of chlorophyll, an important participant in photosynthesis. Iron is essential to the hemoglobin molecule, which is formed by plants as well as animals, where it helps transport oxygen needed for cellular respiration.

Pre-made solutions are the easiest to work with to supply all the needed elements. As with any compound, dosage is important. For very young plants, such as small cuttings or those that are just germinating, 1/3 teaspoon of calcium nitrate dissolved in a gallon of water is about right, for example. Plants that are flowering will require more, about 3/4 teaspoon.

Water and temperature conditions are important factors, as well, when feeding your plants. Any solution should be fed at room temperature, which should also be the temperature of any water used in hydroponic gardens.

Dry plants should not be fed nutrients. Nitrogen burning is possible. That's rarely a problem with hydroponics, but one 'branch' known as aeroponics, where the plants are grown in air, can suffer that problem.

Allowing any water to stand overnight will help evaporate any excess chlorine from home water sources. Mineralized water is preferable to distilled since it will contain calcium and other useful elements. Regulate the pH to keep it as near neutral as possible. As plants take up nutrients they'll tend to make the water alkaline. Add tiny small amounts of sulfuric acid to move it back to neutral. Sodium hydroxide will help shift excessively acidic water back to a neutral pH. Testing kits are available to accurately measure the pH of your hydroponic water.

Hydroponically grown plants are more sensitive to nutrient levels and less able to self-regulate than those in soil-based gardens.

In soil, for example, they can take up or shed compounds. Releasing compounds into the water medium doesn't move them away from the plant. The hydroponic gardener will need to exercise more care to keep plants healthy.

PACIFIC HYDROPONICS

Every day of the year, lettuces from Pacific Hydroponics fly all over the world, not as packaged exports, but dished up as salads and garnishes in thousands of airline meals. As supplier to Qantas and Ansett airlines, Pacific Hydroponics has carved out a secure, but highly particular, market niche which demands year round supply and an assurance of top quality.

Looking at the streamlined growing operation at Pacific Hydroponics today, it's hard to imagine the problems which plagued its creation. But in reality, a period of "no crops and no income" was a part of the difficult birth of today's prosperous enterprise. For owner Garry Cahill, unraveling the mystery, and sometimes secrecy, surrounding commercial hydroponics has been one of the biggest challenges to turning the operation around. Fortunately, persistence has paid off.

Like many of today's hydroponic operators, Garry came into the industry looking for a new business venture.

In 1989 he bought 25 acres of land at Lake Munmorah on the Central Coast of NSW, did a TAFE course in hydroponics, and engaged a company in Melbourne to install a hydroponic lettuce growing operation for him. The company offered a complete package which included the installation of a reliable growing system and an established marketing arrangement with a large supermarket chain. It sounded ideal, but unfortunately the many promises proved hollow. Having paid for the system, Garry then found that the manufacturers were unable to get it going. At the same time, the so called 'established market" failed to materialise. There followed a year with nothing growing in the system and a high interest debt to service.

"In other words," Garry says, "we couldn't get the thing going and we had no market."

Leaving his wife and four children in Sydney, Garry moved up to the farm and spent two difficult years turning the system around and finding a market for his product. His task was assisted by the support of farm manager, Phil

Ritchie, as well as significant contributions from Geoff Creswell of the NSW Department of Agriculture, among others. Between them, they were able to overcome the inherent problems in the system and start producing good lettuce crops. Simultaneously, Garry tackled the question of marketing, convinced of the need for a planned approach.

"I knew we had to get out and market it professionally," he says.

And market it professionally he did, establishing not only an airline market (Qantas also supply meals to all the other airlines), but also a local run to quality restaurants on the central coast. Today, Pacific Hydroponics harvest lettuce every day of the year, delivering them direct to the airlines in Sydney each morning. The farm's distance from Sydney (about 110km) makes this system practical, but Garry comments that he would not go any further away from a capital city market. The farm also benefits from a benign coastal climate, where frosts are rare, and an immediate environment which is conducive to good production. There is good natural windbreak protection, to the west, north and east, and during summer the prevailing north easterly winds pass over the large water area of near by Lake Munmorah, creating a cooling effect. Garry is convinced that a farm's surrounding environment can play a significant role.

"We didn't know that when we built it," he says, "it's a bonus. But now I realise how important environment is to the operation."

At any one time, upwards of 90,000 lettuces are growing in the system at Pacific Hydroponics. Varieties include mignonette, butter, red and green coral, red and green oak, mini cos, green cos and radicchio. Plants are harvested every morning, between 6 and 9am, and around 1000 boxes a week are sold fresh. "Boxes' is actually a misnomer, since Garry has managed to avoid the expense of buying and printing cardboard boxes for packing and transport. Instead, lettuce are packed in red plastic crates which are subsequently returned to him by Qantas for re-use.

Despite the problems Garry experienced in getting to production stage, the system as it stands now is impres sive in its efficiency and ease of operation. It is based on a highly interesting concept, using plastic covered, 14 metre long tables. Seedlings are planted through slits in the plastic cover, which also supports them, and fed by the Nutrient Film Technique (M). The plastic is reflective white on the outside and black underneath, to create a darkened chamber which is favourable to the roots, but not to the growth of algae. There are 410 lettuces in each table, and harvesting involves simply pulling out the plants, roots and all. Hosing down the tops of the tables after harvest is the simple clean up procedure.

The tables are arranged in six sheds', roofiess structures with sides Of mesh fabric that can be rolled up or down as required. Each table, however, has its own plastic cover which can be rolled over to create a mini green house, or to provide hail protection. The sheds measure 110m by 17m and each contains

40 tables. The tables themselves are custom made from fibreglass with slight corrugations, creating small channels down which the nutrient solution flows. The plant roots are free to wander into the nearest channel, and there is no ponding effect. The nutrient is recirculated and the tables have a slope of 300mm over their 14 metre length which, Garry comments, is higher than the norm.

"It's far greater than farms I've seen, " he says, " but I've found out that the quicker you turn the water around the better."

Indeed speed is one of the keys to the way Garry runs his operation.

"We're in the game of fast growing produce – that's what it is," Garry explains. "The faster you grow it, the more money you make because the more crops you get per year. So it is all speed and in fact they're easier to manage the faster they're grown."

The farm uses town water, which is stored in two 45,000 litre (10,000 gal Ion) tanks. It requires pH stabilization before using, since according to Garry, the pH can vary widely – anywhere from pH 5 to 9. The nutrient solution is mixed in smaller 8000 litre (1758 gallons) tanks, which can be heated if - necessary, and a standard lettuce mix nutrient is used, manufactured by Sydney company Simple Grow. Brian Heame, Simple Grow's owner and manager, has been involved with the operation from the start, helping Garry to get his nutrient formulation right. The growing system is a re-circulating one with a flow rate of 1.25 litres per minute and the nutrient solution is topped up when necessary, notably after periods of rain. Each of the six sheds operates quite independently of the others, to minimize production loss should a problem arise. During summer, the solution is dumped about every two months, but dumping is rarely found to be necessary in winter. Interestingly, all monitoring of EC and pH is done manually, using standard meters, and automation is kept to a minimum. This, Garry believes, ensures a more hands-on approach to the operation.

"I really don't think you can fully automate a commerdal set-up. Once you do that, you don't have people here and if you don't have people here, you'll never find out if some thing's wrong. It can't be a 9 to 5 thing," he says. With select clients such as airline companies, high priority has to be given to producing crops that are dean and free from insects. Pests and diseases are minimised by keeping the area around the farm clean, and apart from an attack of Rutherglen Bug two years ago, there have been few problems. Garry subscribes to a preventative approach to such problems.

"We know now, during certain times of the year, you've got to watch out for certain things," he says. "So before it happens we take precautions".

A precautionary approach also extends to the prevention of pythium, which, during the last summer season, didn't manifest itself. Farm manager Phil Ritchie believes management practices have an important influence on pythium prevention and that by not stressing the plants, you can avoid the problem appearing. To this end, he shaded the plants lightly last summer, by letting the plastic covers hang down a little over the tables and believes this had a positive effect. As a result, he is now considering trialing shadecloth as a roof over some of the sheds.

Another pythium preventative measure at Pacific Hydroponics is the addition of Agral, a non-ionic wetting agent, to the nutrient tank. Agral was originally developed in the UK to con trol Big Vein virus in lettuce and has a special registration for use in hydroponic nutrient solutions. It works by dissolving away the tail of the pythium spores, where the virus is alive, and also dissolving the vessicle from which the pythium spores are released. Agral was recommended to Phil by the NSW Department of Agriculture and is now run in the nutrient all summer. No seedling propagation is done on the farm. All the lettuce seedlings come from Leppington Seedlings, professional growers south west of Sydney, as part of a standard weekly order. According to farm manager Phil Ritchie, the relationship with a seedling grower is an important one. "I've explained to him what we've been after in a crop," Phil says. "I can say to him that this type of lettuce is having certain difficulties and then we get a replacement. And he has been able to supply us with lettuces that suit this district."

When Garry Cahill bought the farm, it was being used as a small flower growing enterprise and with it came plantings of Kangaroo Paw (Anigozanthos sp.) and Proteas. These he has maintained as a sideline enterprise, selling the cut flowers on the export market. While Garry admits that he would not have chosen to plant them, he believes in taking advantage of all possible sources of income, particularly since he has a permanent labour force of six people. The flowers are drip irrigated, and the spent nutrient from the lettuce system is also run onto them, rather than dumping it into a paddock. "If we were going to build another of these, we'd make some changes and do it a lot cheaper..."

While Garry would consider diversifying into other crops, he maintains that it would have to be under certain terms. "We'd want a market first and we'd say to the client, OK we'll grow that crop for you and this is how we can grow." The best market, he adds, is a reliable year-round one, rather than one I that fluctuates seasonally. With the benefit of hindsight, Garry can also see ways of improving on his current growing system. "If we were going to build another of these, we'd make some changes and we'd do it a lot cheaper," he says. "So consequently, we'd get it running as well as this is running, but the capital cost would be lower, so we'd be more profitable."

Location is something Garry would not change significantly. He believes there are advantages in being close to the coast, because of the mild climatic influences and consequently would not consider going any further inland, where the frost belt begins.

Out of his adversity, Garry has managed not only to survive, but to turn the enterprise around to profit. His problems, which started with an inept company who were unable to get their own system design up and running, were not alleviated by what he felt at the time was a degree of secrecy within the industry. While things are improving now, Garry feels that a more professional attitude to hydroponic farming operations needs to be adopted, whereby they are run along strict business guidelines.

GREENHOUSE CROPS

The primary crops grown in greenhouses include: pepper, tomato, cucumber, lettuce, herbs, and strawberry. The industry in Florida has changed from primarily either tomato or cucumber in the early 1990s to the diversity of today. A variety of structure types are used, as well. Structures include both fan and pad or naturally ventilated systems. Both are successfully used in the state depending on cropping intentions.

SUBIRRIGATION FOR GREENHOUSE CROPS

Subirrigation is becoming an increasingly common way of watering and fertilizing greenhouse crops in Massachusetts. This article is for growers considering a subirrigation system or just starting out with a new system.

Advantages to Subirrigation

There are three major economic advantages to subirrigation. The most commonly cited advantage is the savings in labour needed for watering the plants: a single person can water thousands of plants by operating the flooding system manually or with the help of a computer. Additionally, there is a potential savings in water and fertilizer with subirrigation since both arc rccirculated and not lost by leaching or run-off. Also, depending on the system and how it is installed, a grower can expect an increase greenhouse space efficiency (percentage of total floor area in use for growing plants). Many growers report more uniform plant growth and less foliar disease with subirrigation. The increase in plant uniformity may be the result of more even and complete moistening of the growth medium and better distribution of nutrients absorbed by capillary flow. The absence of water on the leaves with subirrigation probably results in less foliar disease. The elimination of fertilizer and pesticide leaching and run-off from the greenhouse is a very important reason for using subirrigation. In order to achieve the goal of reduced leaching and run-off the system must be maintained as a truly closed system. The immediate practical value of preventing irrigation effluent from escaping the greenhouse is not always apparent, but protection of water used for drinking and recreation from contamination is probably the most important long-term benefit of subirrigation.

Challenges to Using Subirrigation

Like any other new way of growing greenhouse crops there are a number of challenges to overcome to use subirrigation successfully. The two greatest challenges for most growers is the initial cost of the system and the ability to retrofit the system in an existing greenhouse. A conservative estimate of payback time is 5-10 years, but the period could be as short as 2-3 years depending on the system chosen, whether existing bench frames can be retrofitted, and whether productivity of the system is maintained at a high level. An excellent economic analysis of subirrigation systems was recently published by Wen-fei Uva and her colleagues of Cornell University. Some readers may have heard Wen-fei speak on her work at the New England Greenhouse Conference in October 2000. Her article is very detailed, but concise, and would help growers in choosing a subirrigation system.

Fig. Greenhouse Crops

A grower beginning to use subirrigation will have to learn some new ways of irrigating and fertilizing to use the system successfully. Growth medium and irrigation solution testing for pH and EC is one important skill to acquire. Since the growth medium tends to accumulate salts with subirrigation it is critical to be able to test for EC on a regular basis without having to wait for results from a commercial lab. Also, growers who maintain nutrient and pH levels in the irrigation solution by adding fertilizer or water to stock tanks manually rather than with automatic equipment need to carefully monitor EC and pH to maintain the proper ranges. Successful use of subirrigation requires extra attention to cleanliness to avoid disease and insect problems. The use of pesticides and other chemicals, particularly as drenches, can be problematic with subirrigation so adoption of IPM techniques, especially pest population monitoring, is very important. Cleanliness will be discussed a little more later in the article.

Subirrigation Systems

There are three basic closed, recirculating subirrigation systems currently in use in New England: ebb-and-flow benches, trough benches, and flooded floor systems. There are some variants on these, for example, the "Dutch movable tray system" is very similar to ebb-and-flow, but a complete system is highly mechanized for a number of tasks. Capillary mats and collection trays are also a form of subirrigation, but they are not normally closed systems.

Ebb-and-flow. The ebb-and-flow system is very common and is quite familiar to most growers. The system consists of a shallow, molded plastic bench top which is flooded to water and fertilize the plants; when irrigation is complete the remaining solution drains from the bench and is pumped back to a storage tank.

Ebb-and-flow is very versatile because the bench tops can accommodate all sizes of pots and bedding plant flats (although not on the same bench or irrigation zone at the same time because of the differences in water absorption rates between container sizes).

The bench tops can be installed on existing frames and, with the rolling feature, ebb-and-flow benching can be 80-90 per cent space efficient. Ebb-and-flow benches are easy to retrofit in clearspan greenhouses, but not in greenhouses with many internal supports. This system has the highest initial cost, $4 to 6/ft2, installed on existing bench frames and including tanks, delivery and return pumps, plumbing, and installation. A major portion of the cost comes from the specially molded plastic bench tops which cost about $2.50/ft2. Troughs. This system works by running a film of irrigation solution down a slightly inclined, shallow metal trough holding the plants. The troughs empty in a return channel for recirculation. The pots or flats in the trough have plenty of opportunity to absorb solution as it runs past.

The trough system is very easy to retrofit on existing bench frames. The troughs can be obtained in various lengths and widths from a commercial manufacturer or they can be fabricated by a local metalworking firm to the growers specs. A trough system is about 70-80 per cent space efficient, less than ebb-and-flow, because normally spaces are left between the troughs. Most growers use this system mainly for potted crops, but it is possible to do bedding plant flats if the open mesh style of tray is used to hold the paks. However, because of the trough spacing, it isn't possible to space flat-to-flat except in an individual trough. The initial cost of the trough system is about 2-6/ft2.

The cost of this system can be fairly low if the troughs are made locally or if they are installed on existing benches. Most of the plumbing is simple to put together and inexpensive. Flooded floor. In this system the entire floor of the greenhouse is covered with a concrete carefully designed and installed to pitch towards openings in the floor. Through these openings the irrigation solution enters to flood the floor and, following flooding, the excess drains back to the storage tank. The floors can be installed with bottom heating and divided into zones for separate flooding and bottom heating. Flooded floors can be used to grow plants in all container types and sizes as long as separate irrigation zones are provided for each type.

Space efficiency is about 85-95 per cent. Most greenhouses with flooded floors were built with them rather than retrofitted later. The bottom heating option an efficient way of providing the proper growing temperature for the plants because the air close to the plants is heated and the larger volume of the greenhouse does not have to be heated so much. Some growers complain that in a flood floor plants close to the flood/drain openings tend to be overwatered, especially bedding plants. Also, as in the case of any floor growing system, all the bending and squatting needed to work with the plants can be tiring for workers. Initial cost for a flooded floor is $3-5/ft2, but costs can vary significantly depending on the amount of excavation required for the storage tanks and piping, whether or not bottom heat is installed, and whether the floor is divided into zones for separate irrigation. A very skilled concrete contractor is needed to get the pitch of the floor right to encourage proper drainage and to prevent puddling.

FERTILIZING SUBIRRIGATED PLANTS

Since there is little or no nutrient leaching with subirrigation less fertilizer is needed compared to traditional overhead watering. The general rule for fertilizing subirrigated plants is to use one-half the rate (ppm) of fertilizer normally applied by overhead irrigation. Several years ago I subirrigated poinsettias with solutions of 100, 175, 250, or 325 ppm N from peat-lite 20-10-20 fertilizer. The plants finished about the same size with nearly as large bracts as plants watered from overhead. Leaf analysis revealed normal levels of most nutrients at all fertilizer rates and no evidence of a serious nutrient deficiency or excess. EC (soluble salts) levels were higher with subirrigation than overhead watering. EC was highest near the top of the growth medium because of surface evaporation and deposition of nutrient residues. None of the treatments developed had excess EC.

The results of this study demonstrated that poinsettias grow well over a wide range of fertilizer concentrations in subirrigation including levels applied by traditional overhead watering. In fact, most growers I've visited in New England who subirrigate poinsettias on a large scale use fertilizer rates in the range of 200-250 ppm N. Use of fertilizer rates above 250 for subirrigated poinsettias increases the risk of excess EC leading to growth inhibition and plant injury. Learning to use an EC meter to monitor soluble salts on a regular basis is very important with subirrigation.

Chemicals and Subirrigation

Many insect and disease problems can be prevented by adopting a new standard of greenhouse cleanliness and through the use of simple IPM practices to prevent infestations and infections from getting out of control. This means that for now growers must apply pesticides as they would to overhead watered plants only more carefully. Heavy or frequent foliar spraying, or use of growth medium drench treatments, are risky practices because enough chemical may enter the irrigation solution to cause undesirable effects to the plants in the long term. To avoid this problem, some growers divert irrigation water from their subirrigation system for conventional disposal following a pesticide application rather than letting it return it to the tank for recirculation. In the absence of definitive information on the extent of buildup and effects of recirculated chemicals, growers should try to limit pesticide treatments as much as possible especially growth medium drenches.

Zero Tolerance™ disinfectant is one chemical that can be recirculated in subirrigation with beneficial effects. Zero Tolerance™ can control algae and a wide variety of root disease organisms. The product label has specific directions on its use in subirrigation systems.

Interestingly, there is some interest in applying plant growth regulators (PGRs) through subirrigation. Currently A-Rest™ and Bonzi® are labeled for use in "chemigation" systems including subirrigation by ebb-and-flow and from saucers. Labels for both PGRs have detailed instructions on how to apply the chemical so as not to cause plant injury and to protect water supplies. In the author's opinion, it too early to draw conclusions about the efficacy and safety of PGR application this way but it is being studied in Florida and I will have some preliminary results to report soon.

Finally, cleanliness is very important. As a routine practice dead plant material and other large "stuff" should be removed from growing areas, inside tanks, and plumbing after each crop. Then the system should be disinfected with Zero Tolerance™ or Green-Shield™.

SUCCESSFUL CROP PRODUCTION IN COMMERCIAL GREENHOUSES

Successful crop production requires that crop pests and diseases be managed so that the effects of diseases and pests on the plants are minimized. The management of crop diseases is directed at preventing the establishment of diseases and minimizing the development and spread of any diseases that become established in the crop.

Managing pest problems is directed at preventing the pest populations from becoming too large and uncontrollable. The presence of pests and diseases are a fact of crop production and growers must use all available options and strategies to avoid serious pest and disease problems. Integrated pest management (IPM) is a term used to describe an evolving process where cultural, biological, and chemical controls are included in a holistic approach of pest and disease control. Key components of effective pest and disease control programmes include: crop monitoring, cultural control, resistant cultivars, biological control and chemical control.

CROP MONITORING

Crop monitoring is the continually on-going surveillance to detect the presence of a pest or disease at the very early stages of development of the disease or pest population, before economic damage has occurred. Everyone involved in working the crop should be made aware of the common pest and disease problems and what to look for to detect the presence of problems in the crop. In addition this general surveillance of the crop, dedicated monitoring of the crop should be included in the weekly work schedule. Blue sticky cards, placed throughout the crop, are a useful monitoring tool to help trap and detect pest problems before they become a problem.

Yellow sticky cards are known to attract and catch some biological control agents *e.g. Aphidius sp*. (Don Elliott, pers comm) Biological control agents can be released well in advance of any pest population explosion thus allowing for the establishment of the control agents and prevention of a serious pest problem. Crop monitoring should begin when the crop is still on the seedling table or at the transplant stage (especially when transplants are obtained by another greenhouse *i.e.*. Purchased from a propagator). If transplants are being purchased from a propagator it is important to be in contact with the propagator regarding any pest problems encountered during the production of the transplants.

It is also important to know what pest control measures were used, if any, to control the problems. It is advisable to establish in advance, what control measures you are willing to have applied to the transplants at the propagator's prior to receiving them into the greenhouse. The concern is that any pesticides that are applied are compatible with the pest control programmes *i.e.* biological control programmes that will be used for the duration of the crop. Some growers may insist that only biological controls be used during the production of transplants, and/or that biological control agents be introduced preventively to the transplants before they are received.

Cultural Control

Cultural control involves providing the conditions that favour the growth, development and health of the crop, and where ever possible, providing conditions that work against pest and disease. Many disease causing fungi and bacteria require the presence of free water or condensation on the plants in order to cause disease. High relative humidity promotes the development of disease, and maintaining the environment below 85 per cent relative humidity will help to escape or avoid disease problems. Ensuring proper ventilation and air movement within the crop canopy, as well as maintaining optimum plant spacing and a relatively open canopy, will ensure good air circulation and minimize the establishment of micro-climates that favour disease development. Proper contouring of the greenhouse floor will avoid the pooling of water which contributes to localized high relative humidity. Optimizing the greenhouse environment to favour the development of the plant will ensure a strong, healthy plant which is not only a prerequisite for high yields but also results in plants that are better able to resist diseases and insect pests.

Good crop sanitation is another important component of successful cultural control. The plants must be pruned and maintained on schedule, all crop debris should be promptly removed from the vicinity of the greenhouse. Any weeds that happen to gain a foot hold through gaps in the floor plastic should be removed immediately upon discovery and the floor repaired. Personal plants "pet" plants should not be grown in the greenhouse. Both weeds and "pet" plants can be as source and "haven" for pest and disease problems.

Pruning tools and other equipment should be cleaned and disinfected on a regular basis. Aprons or other clothing worn by the workers should be washed frequently. When a disease or pest problem area exists in the greenhouse, that area of the greenhouse should be worked last, to avoid the spread of the disease or pests by the workers. In this situation, special care must be taken to disinfect tools and to clean clothing. Maintain a 6 to 10 meter wide buffer zone around the outside of the greenhouse by regularly mowing any weeds that try to grow in this zone. The presence of plants in close proximity to the greenhouse can serve as a reservoirs for continual introduction of pests and diseases into the greenhouse. Screening of the intake vents can also play an important role in excluding pests from the greenhouse. It is not enough just to screen-off the intake vents as the screening restricts the air flow into the greenhouse, it is important to ensure that the surface area of the screening used is large enough so that it does not restrict the flow of air into the intake vents. This may require that a screen 'chamber' be constructed over the vent.

Resistant Cultivars

Plant breeders have had considerable success in developing cultivars that contain genetic resistance or tolerance to diseases. When selecting the cultivars to be grown, it is important to consider the genetic resistance of the cultivars to the prevalent disease problems in the region. The development of cultivars possessing genetic resistance to pests has been relatively unsuccessful, however, the techniques of genetic engineering have made inroads in conferring pest resistance in plants. Genetically modified, pest resistant plants may become available to greenhouse growers in future. The development use of genetically modified plants or genetically modified organisms (GMOs) is currently a contentious issue, and may not be accepted by growers or consumers.

Biological Control

Biological control uses beneficial organisms, primarily predators and parasites, to control pest populations below economically important levels. The goal is to establish a balance between the pest population and its parasites and predators to keep the pest population under control. The complete eradication of the pest population is not the goal of biological control programmes, as some pest organisms are required so the parasites and predators can reproduce.

The greenhouse industry has a well established reputation for using biological pest control agents more than any other crop production industry. The reason for this is, in part, due to the ability of greenhouse growers to manage the environment to favour the biological control agents. Another factor is the relatively limited number of pest species in greenhouses, as well as a general tolerance of greenhouse crops to leaf damage caused by these pests. The high value of greenhouse produce is another reason why the use of biological controls is economical in greenhouse crops. The increased use of biological controls has led to a reduction in pesticide applications as the industry leads in environmentally responsible, intensive crop production.

Effective biological control of diseases is a more difficult goal and to-date, has rarely been achieved. However, research in developing biological controls for greenhouse crop diseases is ongoing and it is likely that biological control products for greenhouse diseases will be available in Canada in the near future. The primary strategy of biological control for greenhouse plant diseases is to introduce fungal parasites to control populations of disease causing fungi in the greenhouse environment so that they are unable, or have a reduced ability to infect the plants. Some of the promising biological control agents, for example, fungi in the Genus *Trichoderma* are also strong competitors of the disease causing fungi such as *Botrytis cinerea*, and can be used to protect wound sites to prevent *Botrytis* from colonizing the wound site.

Chemical Control

Pesticides are valuable tools when used as a component of an integrated pest management programme. Insecticides should be applied only in support of biological control programmes, dealing with localized pest outbreaks in the crop that have escaped the biological control agents. When insecticides are used, care must be taken to ensure that they are compatible with biological control agents, that there will be minimal long term adverse residual effects on biological control programmes. Fungicides are used only when a disease problem is detected.

Pesticides are regarded as the controls of last resort because their misuses creates high-profile environmental and food safety problems. Also, the application of some pesticides to a crop can cause stresses that reduce the productive life of the crop and can make the plants susceptible to other pests and diseases. If the use of biological control agents is to obtain a balance between pests and predators that does not threaten the productive yield of the crop, the indiscriminate use of pesticides creates imbalance and uncertainty in the crop.

GREENHOUSE ENVIRONMENT TO CONTROL PLANT DISEASES

Numerous plant disease problems can arise in greenhouse situations. These diseases can cause extensive damage of allowed to develop unchecked. Since plant disease are strongly affected by temperature and humidity, the best way to combat disease is to manipulate the greenhouse environment. Unlike the weather outdoors, we can control the greenhouse environment.

Plant disease control in the greenhouse is generally more effective if the following aspects of the greenhouse environment are managed properly:

HUMIDITY

High humidity levels encourage the development of many plant diseases. The relative humidity is usually 25 per cent-70 per cent during the day in greenhouses and generally no problem. However, humidity levels are generally 90-100 per cent during the night. During periods of rainy weather in winter, the relative humidity may stay near 100 per cent for a number of days and nights. Maintain adequate plant spacing. When plants are crowded together, disease development is encouraged by the high humidity in the canopy. Plants hung overhead reduce normal water evaporation and contribute to high humidity in the crop canopy.

Maintain air circulation during periods of high humidity. Most greenhouses are equipped with air circulation, fan-jet or horizontal air flow, systems. These systems should operate continuously when high humidity occurs in the greenhouse, *i.e.*, every night of the year and during rainy overcast days. Ventilate the greenhouse to reduce internal relative humidity. Most winter evenings are cool enough to raise the humidity to 100 per cent and cause considerable condensation in the greenhouse. This condensation can be reduced if the greenhouse is ventilated at dusk each day.

In the late afternoon, turn on the ventilation fans to exhaust the warm moist air from the greenhouse. The warm moist greenhouse air is replaced with cool/cold, moist air from outdoors. When this outside air is heated in the greenhouse it becomes much drier than the previous greenhouse air. This management practice greatly reduces the relative humidity in the greenhouse and reduces potential disease problems.

Watering

The risk of plant diseases is reduced when the foliage and flowers are kept dry. Most disease organisms need water on the plant surface for normal growth. Additionally, splashing water is the primary method of spreading disease from plant to plant. Apply water only to the growing medium surface, when possible, rather than "showering" the whole plant. Water early in the day; don't water after 4 p.m., except in the summer. Watch plants closely and water judiciously. Remember that with variations in light levels, temperature and humidity the plants need for water will change. Watering practices should be attentive to these changes. Plants beneath overhead plants will not dry as quickly as neighbouring plants in full sun. Water thoroughly and do not water again until the growing medium is dry. Excess water in the growing medium weakens plants and makes them more susceptible to damping-off and root rotting diseases. Good drainage and the related good aeration of the growing medium is important to prevent root rot problems. Separate weaker plants that do not use water as fast as the strong plants; They will not be over-watered as easily. Be sure to use a good growing medium. Professionally prepared media are generally the most effective and are recommened to commercial growers. When using mineral soil in a mix, be sure it is adequately sterilized. Never use more than 15 per cent mineral soil in a growing media that will be used in containers less than 4 inches tall.

Sanitation

Keep the greenhouse clean and free of plant debris and outside soil. Remove dead leaves, flowers, plant refuse and weeds from the greenhouse. Debris should be gathered regularly and discarded as soon as possible after collecting. Immediate removal is important because certain fungal pathogens can develop and produce spores on the plant debris. Weeds along walkways and under benches or even weeds growing just outside the greenhouse can harbour diseases that can be transmitted to greenhouse crops. Insects on weeds may vector plant virus diseases.

Non-sterile soil from outdoors should not be allowed into the greenhouse. Since many soilborne pathogens can be a serious problem if introduced into the greenhouse, the best control method is to exclude these pathogens from the greenhouse. This can be accomplished by using a soilless medium which is free of plant pathogens. Surface sterilize all work surfaces and tools regularly to prevent accidental disease spread.

Diseases are often transferred in the greenhouse through used pots, dirty tools, messy work surfaces, unswept floors, etc. Precautions should be taken to prevent this. Surface sterilization of tools and work surfaces in and around the greenhouse can be accomplished by washing with a dilute solution of household bleach (9 parts water to 1 part beach) or the use of commercial disinfectants such as Greenshield and Physan.

GREENHOUSE PLANT DISEASES AND THEIR CONTROL

The construction of greenhouses must foster a number of critical conditions for the optimal health of plants. These include good soil drainage, ample sunlight, optimum water and air quality, a good exchange of air with the use of adequate-sized exhaust fans, continuous fresh air and ample air movement between the plants. Given the fact that there are a large number of plants confined in a relatively small space it is common for greenhouse plants to develop viral diseases, leaf spots, root rot and other diseases very quickly.

DAMPING OFF

Damping off is caused in greenhouses by the pathogens Pythium spp., Fusarium spp. and Rhizoctonia solani. The pathogens are likely to infest the seedlings of nearly all plants. The disease is most common due to poorly drained soil in spring; it causes the death of the younger seedlings, and root rot and stunted growth in older seedlings.

Recommended control includes maintaining moderate soil moisture and ensuring proper soil drainage. Seeds should also be germinated at higher temperatures. Chemical fungicides for treating seeds include captan, mefenoxam and metalaxyl.

Black Root Rot

Black root rot is a common disease in floral greenhouse plants and is caused by the fungus Thielaviopsis basicola. Greenhouse plants most susceptible to root rot include geranium, poinsettia, pansy, fuchsia, petunia and begonia. The fungus is found in soil and can live in the soil without causing disease for many years in thick walled spores. The disease causes stunting, yellowing, wilting and eventual death of the entire plant. Management of the disease includes avoiding plant stress by unnecessarily cool or warm temperatures, ensuring well drained soils and correcting any nutritional imbalance in soil.

Gardening procedures should be sanitary, such as avoiding reusing trays and containers and using good quality potting soils. Chemical controls include soil drenching with Domain, Banner and Terragard.

Powdery Mildew

Powdery mildew is a frequently occurring disease in greenhouse plants and is caused by various fungi like Erysiphe, Leveillula, Microsphaera and Spaerotheca. The fungi survive on greenhouse weeds and crops and attack the healthy, vigourous plants. Powdery mildew appears on floral plants like African violet, dahlia, begonia and hydrangea. It is also common on herbaceous perennials like aster, phlox and sedum. Herbs like rosemary, sage, mint and greenhouse cucumbers and tomatoes are also susceptible.

Prevention includes proper spacing of plants to reduce humidity levels, keeping humidity level below 93 per cent, and keeping greenhouse clean of weeds. Since powdery mildew only colonizes the upper cell layers, there are no recommended chemical control methods for the disease.

STRUCTURE OF GREENHOUSE

A greenhouse is a structure with different types of covering materials, such as a glass or plastic roof and frequently glass or plastic walls; it heats up because incoming visible solar radiation (for which the glass is transparent) from the sun is absorbed by plants, soil, and other things inside the building. Air warmed by the heat from hot interior surfaces is retained in the building by the roof and wall. In addition, the warmed structures and plants inside the greenhouse re-radiate some of their thermal energy in the infrared, to which glass is partly opaque, so some of this energy is also trapped inside the glasshouse. However, this latter process is a minor player compared with the former (convective) process. Thus, the primary heating mechanism of a greenhouse is convection. This can be demonstrated by opening a small window near the roof of a greenhouse: the temperature drops considerably.

This principle is the basis of the autovent automatic cooling system. Thus, the glass used for a greenhouse works as a barrier to air flow, and its effect is to trap energy within the greenhouse. The air that is warmed near the ground is prevented from rising indefinitely and flowing away. Although there is some heat loss due to thermal conduction through the glass and other building materials, there is a net increase in energy (and therefore temperature) inside the greenhouse. Greenhouses can be divided into glass greenhouses and plastic greenhouses. Plastics mostly used are PEfilm and multiwall sheet in PC or PMMA. Commercial glass greenhouses are often high tech production facilities for vegetables or flowers. The glass greenhouses are filled with equipment like screening installations, heating, cooling, lighting and may be automatically controlled by a computer.

USES

Greenhouses protect crops from too much heat or cold, shield plants from dust storms and blizzards, and help to keep out pests. Light and temperature control allows greenhouses to turn inarable land into arable land, thereby improving food production in marginal environments. Because greenhouses allow certain crops to be grown throughout the year, greenhouses are increasingly important in the food supply of high latitude countries. One of the largest greenhouse complexes in the world is in Almeria, Spain, where greenhouses cover almost 50,000 acres (200 km^2).

Sometimes called the sea of plastics. Greenhouses are often used for growing flowers, vegetables, fruits, and tobacco plants. Bumblebees are the pollinators of choice for most greenhouse pollination, although other types of bees have been used, as well as artificial pollination. Hydroponics can be used in greenhouses as well to make the most use of the interior space. Besides tobacco, many vegetables and flowers are grown in greenhouses in late winter and early spring, and then transplanted outside as the weather warms. Started plants are usually available for gardeners in farmers' markets at transplanting time. Special greenhouse varieties of certain crops such as tomatoes are generally used for commercial production. The closed environment of a greenhouse has its own unique requirements, compared with outdoor production. Pests and diseases, and extremes of heat and humidity, have to be controlled, and irrigation is necessary to provide water. Significant inputs of heat and light may be required, particularly with winter production of warm-weather vegetables. Because the temperature and humidity of greenhouses must be constantly monitored to ensure optimal conditions, a wireless sensor network can be used to gather data remotely. The data is transmitted to a control location and used to control heating, cooling, and irrigation systems.

GREENHOUSE TECHNOLOGY AND DESIGN

The Dutch and pseudo Dutch (particularly Britain and Sweden) industries are founded on glass. Two key greenhouse designs are used – the narrow glasshouse, often called the Dutch Venlo is the dominant design used for vegetable production, while the wide span is favoured in the production of cutflowers.

The height of greenhouses has a significant impact of the capacity to control the growing environment. In these industries, the older structures that have not been replaced have gutter heights in the range of 3 – 4 metres. There are few of these structures remaining. More recently constructed greenhouses have higher structural designs. Common heights are 4.5 metres, with some being built to 5.5 metres. Some greenhouse manufacturers expect future designs will have higher gutters.

Thermal screens are used in a lot of this industry as a method of saving energy (as much as 19 per cent) in heating the greenhouse, however, though they can have a dual purpose by preventing excess energy from entering the greenhouse in summer, they are seldom used for this second purpose. Growers have found that when drawn, the adverse impact on ventilation, coupled with lower light levels is not acceptable. Subsequently, greenhouses are whitewashed for the summer period instead. Many operators apply the whitewash through sprinklers installed on the roof of the greenhouse using products that either readily wash off or becomes clear in rain. This ensures maximum light reaches the crop in overcast conditions.

The Dutch industry produces approximately 60 per cent of the cutflowers sold in the world market and 50 per cent of potted plants. The production of potted plants is increasing at a faster rate than cutflowers. In vegetables, tomatoes have the reputation as the largest crop area, but new developments have tended towards capsicums. Overseas, growers build greenhouses for particular crops; they specialise and become expert in a single crop. Very few growers produce more than one type of crop.

Automation of the greenhouse is standard practice in these industries. As in Australia, labour is expensive and often difficult to recruit. Computers monitor and control the growing environment as well as the nutrient dosing and irrigation of the hydroponics system. The pinnacle of this drive for automation is found in some of the nursery operations in which a computer is able to control not only the growing environment of the plant, but also move and sort plants within the greenhouse. In one 20 hectare nursery, a total of just six employees, are involved in little more than supplying inputs such as potting mix to the potting machine and nutrients to the dosing system, and plastic sleeving the plants that are ready for shipping –everything else is undertaken by the computer.

THE CANADIAN INDUSTRY

The greenhouses used in Canada are similar to the Dutch industry and the Dutch influence is quite strong. Growers travel to The Netherlands on a frequent basis. The primary difference is seen in the use of plastic cladding, which is rarely used in Holland. Plastic offers the Canadians a less expensive option and because of the higher light levels experienced in Canada, particularly Ontario, compared with Holland, there are not the same disadvantages. As a result, plastic greenhouses outnumber glass by roughly 2:1. A notable innovation, however, is the development of curved glass sheets enabling the construction of glasshouses with curved roof designs. A curved roof permits a higher average level of light transmission over the course of the day and season.

Structures in this industry have high gutter heights, which enable improved management of the growing environment.

Greenhouses tend to have gutters 4 – 5 metres above the ground level. Older sheds with smaller air volumes (lower rooves) are generally used for cucumber production because they are less suitable for solanaceous (tomato, capsicum and eggplant) crops.

The floricultural industry makes greater use of glasshouses. The largest flower producing area in the country is the Niagara Peninsula. In British Columbia floriculture has been growing at a rate of around five per cent per annum. This is similar to most other countries, including Australia, which are experiencing a strong demand for cutflower crops. Fully automated floriculture greenhouse operations may cost as much as \$350/m2 to build.

The Spanish Industry

There are over 40000 hectares of greenhouse structures in Spain, concentrated along the coast from Alicante on the east to Almeria in the south. The majority of the structures are of the traditional Spanish greenhouse, but new developments are mostly of modern European designs. Almost half this area is used to produce tomatoes, though flower exports from Spain are growing annually.

The structures in the traditional Spanish industry are basic. Typically a tent structure, the greenhouses are used for production of vegetable crops sold into the European supermarkets. The supermarket chains tend to source Spanish product during winter. This traditional greenhouse evolved in Spain as poor land holders began to realise the benefits of covering crops with plastic. These structures are generally unheated and poorly ventilated. The greenhouses are used to produce crops from autumn to spring. Very few are used during the hot summer months. The structures are essentially two layers of wire netting suspended by posts. The older greenhouses have wooden posts, while the more recently built structures make use of stainless steel. Plastic film and occasionally woven plastic sheeting is inserted between the netting. Production may be in either hydroponics or soil. Though the move towards soilless systems is occurring, the vast scale of the industry and the relaxed attitude of the growers will probably mean that soil production will continue for many years to come.

The newer greenhouse industry takes after the more typical European, that is Dutch, designs. Structures have a minimum gutter height of 4 metres, several extending to 5.5 metres to attain better environmental management during the intensely hot summers.

Plastic and glass cladding are used, though where the investment directly involves a Dutch connection, glass is used without question – a habit more than a requirement, as the Spanish industry is quite south compared with other greenhouse production areas in Europe and consequently receives a lot more sunlight.

Polycarbonate is increasingly popular and is used on some or all the walls as this material provides extra structural integrity in the face of moderate sea breezes. Gable and curved roof designs are most common. Hydronic heating is standard and often involves co-generation of carbon dioxide for use in the crop. Insect screens are used in many of the modern structures.

WATER DISINFECTION AND REUSE IN GREENHOUSE HORTICULTURE

The production in European greenhouse horticulture is not yet as water efficient as it could be. Soil-less growing systems are becoming common in horticultural practice in most of the European countries, although not in each country on a large scale yet. The advantages of soil-less growing systems compared to soil grown crops are as follows:

- Growth and yield are independent of the soil type of the cultivated area;
- Better control of growth by use of improved water quality and a better fertilization;
- Increased quality of products;
- Pathogen-free start by use of substrates other than soil and/or easier control of soil-borne pathogens.

The disadvantages of these systems are:

- The required high quality of water;
- High investments and high costs for fertilizers;
- Low quantity of the water.

In most cases open or run-to-waste systems are adopted. In such open systems, superfluous nutrient solution freely leaches to ground and surface water. Because of economical motives and environmental concerns closed soil less systems can be applied. These closed systems are more efficient with the use of water and fertilizers, and cause less damage to the environment. The disadvantage of the closed systems is the risk of a rapid dispersal of soil-borne pathogens by the recirculating nutrient solution. To eliminate these pathogens, several disinfection methods can be used:

OZONE TREATMENT

Ozone treatment can be used to disinfect the drain water. Ozone is the second most powerful sterilant in the world and its function is to destroy bacteria, viruses and odours. An ozone supply of 10 g/h/m3 water with an exposure time of 1 h is sufficient to kill all pathogens.

UV Disinfection

Another way to disinfect the drain water is the use of UV-radiation. Ultraviolet radiation (or UV) is a proven process for disinfecting water, air or solid surfaces that are microbiologically contaminated. For eliminating bacteria and fungi an energy dose is recommended of 100mJ/cm^2. For viruses a dose of 250 mJ/cm^2 is recommended.

Heat Treatment

When heat treatment is applied, a solution is heated for about 30 seconds to a temperature of 95ÚC. At this temperature all pathogens are killed. A disadvantage of heat treatment is the consumption of gas. Also warm drain water contains less oxygen.

Slow sand Filtration

For several years commercial growers have used a slow sand filtration installation to eliminate pathogens.

Sand filtration is frequently used and a very robust method to remove suspended solids from water. The filtration medium consists of a multiple layer of sand with a variety in size and specific gravity. Sand filters can be supplied in different sizes and materials both hand operated or fully automatically.

IRRIGATION WATER QUALITY FOR GREENHOUSE PRODUCTION

Most greenhouse media contain 30 to 60 per cent peat moss alone or in combination with composted pine bark. Other materials are added for drainage and aeration. In terms of air/water relations in the root zone, the quality of the peat used is very important. Peat that has been milled too much has a smaller fibre size. Media settling may result in loss of plant-rooting volume. Also, aggregates such as vermiculite may or may not improve drainage and air space, depending on the size and shape of the particle. Fine-grade vermiculite particles may actually decrease media aeration. Only a portion of the water added to media is available for root uptake. Available water-holding capacity is the amount of water held in the root zone and available to plants between irrigating and when the plant wilts. In a 6-inch pot, approximately 65 per cent of the pore space is filled with water after the pot has been saturated and allowed to drain.

Generally only about 70 per cent of that water is available; the rest is called unavailable water. The amount of available water depends on how tightly the water is held to the particles of materials that make up the media (matric tension).

For example, peat has relatively higher unavailable water contents at a given matric tension compared to rock wool. This variability in the availability of water in different types of media components means no two media are exactly alike in terms of providing water to plants. This makes knowing when to water difficult. Another important characteristic of media components that influences watering practices is wettability, *i.e.*, the ability of dry media to rapidly absorb water when moistened. The choice of media should be influenced by irrigation systems and practices.

MEDIA COLUMN HEIGHT/CONTAINERS

Another factor relating media to air/water relations in the root zone is the size of the growing container. With media in containers, the amount of air and water held in a given media is a function of the height of the column of media. The taller the column, the smaller the ratio of water to air spaces.

This is most important in plug production where the small cells drain very poorly or not at all, resulting in poor root zone aeration.

The dramatic effect of container height on air space is evident, which shows the change in per cent volume air space of a 1:1 peat:vermiculite media in various growing containers.

In all containers, there will be a certain amount of saturated media at the bottom of the container after drainage. This is due to what is called a >perched water table. The saturation zone is a larger part of the total volume of the growing media in a very short container, such as a plug cell. A good way to illustrate the effect of container height is to use a sponge. A sponge of the dimensions 2@ x 4.25@ x 8.5@ (72.25 cubic inches or 1,184 milliliters) represents the media in a container. When fully saturated, the sponge holds 950 ml; that is, the total porosity is 80 per cent. Holding the sponge so it is 2 inches high results in about 50 ml water draining out, resulting in a volume air space of 4.2 per cent. If it then is held so it is 4.25 inches high, another 125 ml drains out, resulting in a volume air space of 14.8 per cent. If the sponge is then held so it is 8.5 inches high, another 375 ml drains out, resulting in a volume air space of 46.5 per cent. Starting with the same volume of media (sponge), the effect of container height (sponge height) on media air space is dramatic. We can conclude that the choice of containers is important in managing water/air relations in the root zone, especially of plugs.

Another issue is whether square or round plug cells are better. In general, square cells with their greater volume make crop management a little easier. As long as the height is the same, however, there is no difference in air space.

Media Handling

How media are handled can greatly influence their air and water characteristics. The major concern is to avoid compaction. Containers, including plug trays, should be lightly filled and the excess brushed off the top. Air space can be drastically reduced by compaction. At no time should any growing containers be stacked. The moisture content of the media prior to filling containers may also be important. Adding water to peat based mixes before filling plug trays causes the media to swell and helps create more aeration. Water added to about 100 per cent by weight of the media is sufficient for cell packs. Plug mixes should have about 200 per cent by weight water added before filling plug trays. Moistening of the media before filling larger containers does not have much benefit.

SAND CULTURE

This term may be used in a general way to refer to cultures in sand, fine gravel or cinders, but here it applies specifically to cultures of plants in sand, using nutrient solutions to supply the mineral elements required for plant growth. Sand is adapted ibr such use where free drainage is possible, as in the constant-drip and flush or "slop" techniques. In small installations, coarser grades of sand may be used with sub-irrigation, which see. In larger installations, it is not suitable for the subirrigation method because of slow drainage, but it can be used in a greenhouse bench with free drainage and the surface application of the nutrient solution. The best type of sand is a quartz sand of the grade used for making concrete. It should be washed clean. This can be done by placing the sand in a pail or tub and forcing water up through it by delivering the water through a hose at the bottom of the receptacle so the flooding will carry out clay and silt particles. Washing is especially important if the sand is fine. If the sand is too coarse, it will require more nutrient solution to maintain the proper moisture relation.

If the only source of sand available is a limestone sand, it may be advisable' to use cinders, for the lime sand contains much calcium which, in excess, will cause plant foliage to turn yellow. However, by using a more acid nutrient solution, or by running the nutrient solution daily through the limestone sand for two weeks before placing the plants in the medium, the sand grains can be given a coating of phosphates and be made safe for growing plants.

GROWING SEEDLINGS IN SAND

Seedlings can be raised by sand culture either to be transferred to cultures for chemical gardening or for use in the outdoor garden. By this method seedlings with very fine root systems can be produced; if handled properly they will transplant into soil readily and, given proper care, start to grow promptly. In addition, losses from damping-off are minimized, while another advantage is that if the seedlings are not too crowded, they can remain in the sand if supplied with nutrient solution in the proper concentration.

A good way to handle them is to construct a box not less than 4 in. Deep on the inside with openings in the bottom so excess solution can drain out. These spaces between the bottom boards or small drainage holes can be covered with cheesecloth, burlap, or glass wool to prevent the sand from sifting through. Fill the box with clean sand and flush it well with water to settle the sand thoroughly. Then flush with nutrient solution diluted with five times as much water. The seeds may be sown in drills or rows or broadcast; if in rows, make the furrows barely deep enough so that the seeds can be covered after being sown thinly. If the broadcast method is used, spread the seeds uniformly and not too thickly over the surface, and cover with about one-fourth inch of sand. In either case, firm or compact the sand, then apply the dilute nutrient solution until the sand is saturated.

Cover the box with newspaper to reduce evaporation and place in the proper temperature for the kind of seed sown. Since it would be fatal to the embryo of the seed if the surface should become dry, the sand must be kept moist. The dilute nutrient solutions should be applied as a fine spray, to prevent disturbing the surface; a rubber bulb syringe with a fine rose nozzle is excellent for this purpose.

As soon as the seeds have germinated and the shoots appeared above the surface, remove the newspaper covering. When the seedlings have grown to suitable size, they may be transplanted to a permanent location or to other vessels where they can be more widely spaced.

If it is desired to leave them in the flat, thin them out to a proper spacing. They are now able to manufacture food in their leaves, and the nutrient solution should be applied in a stronger concentration – one part of nutrient solution to one part of water. Should it be desired or necessary to remove the seedlings from the sand medium, flood it with nutrient solution and gently pull the plants out one by one, assisting the removal with sonic instrument thrust under the root. A label, spatula, or case knife will be suitable.

"Sand culture is one of the most efficient and cost-effective methods of soilless culture, and is widely used in the dry arid regions of the Middle East. Although it is not used in Australia on a commercial scale, it has proved popular among some growers for crop trials. Its simplicity and low capital cost makes it an attractive alternative to existing growing methods."

Sand culture is one of the most efficient and cost-effective methods of soilless culture, and is widely used in the dry arid regions of the Middle East. Although it is not used in Australia on a commercial scale, it has proved popular among some growers for crop trials. Its simplicity and low capital cost makes it an attractive alternative to existing growing methods." by Leo Wright History records that Aztec Indians were among the first to use sand culture techniques.

During the 12th century the Aztecs developed extraordinary irrigation systems in the Mexican basin, with swamp reclamation their most significant achievement, even including the colonisation of surrounding lakes. The Aztecs grew beans and squash on primitive rafts covered with sand removed from shallow lake beds. The plant roots grew through the sand down into the nutrient-rich water of the lake. A few of these so-called 'floating gardens' can still be seen on the lakes surrounding Mexico City today.

In modern times, sand culture re-emerged during the 1960s, ironically in Puerto Penasco, Mexico, where successful trials led to commercial operations in Mexico, the U.S. Southwest and in the Middle East. Using coarse beach sand, leached free of excess salts, vegetables were either sown directly in the sand or planted as seedlings.

The Advantages

The sand culture system has many advantages over traditional hydroponic techniques. The fact that it is an 'open' ('run-to-waste') system, whereby the nutrient solution is not recycled, greatly reduces the likelihood of diseases such as Fusarium and Verticillium spreading in the medium. It also means there is no nutrient imbalance since plants are fed with fresh nutrient solution at each irrigation cycle. Another advantage is the excellent capillary action of sand, which results in lateral movement of nutrients so that there is an even distribution of nutrients throughout the root zone. Additionally, water retention is high owing to the smallness of the sand particles, allowing fewer irrigation cycles during the course of the day. Unlike other systems, particularly NFT (Nutrient Film Technique, in which there is no medium), in the event of a fractured pipe, power or mechanical failure, there is more time available to repair the system before plants consume existing water in the medium and begin to experience stress due to dehydration. Practical advantages of sand culture include lower construction costs, simplicity of operation, and easy maintenance and service.

The Disadvantages

The disadvantages of sand culture are few, and in some cases are in common with other hydroponic techniques. A major disadvantage is the need to use chemical or steam sterilisation between crops in order to destroy media-borne pathogens. Such methods are thorough, although somewhat time-consuming. Like all drip irrigation systems, feed lines can become blocked with fine nutrient particles, grit or sand. This can be overcome by using in-line mesh filters which can be easily cleaned. Perhaps the overriding factor which makes sand culture unattractive to end-users is the seemingly high consumption rate of nutrients because of the need to run to waste.

However, with careful management, the waste should account for no more than 8 per cent to 10 per cent of the total nutrient solution added. Salt build-up is another common problem, but this can be corrected by flushing the medium periodically with fresh water. Again, regular and careful monitoring of the drainage water for evidence of salt accumulation is important to prevent excess salt problems.

Sand Culture Systems There are basically two proven methods of utilising sand as a growing medium. The first is to use plastic-lined beds in above ground troughs; the other is to spread sand over the entire floor of the growing area. Both methods use a sand depth of between 300 and 400 mm. Sand beds can be easily constructed with wooden sides, or with concrete reinforcement wire, cut and bent to form an above ground trough and lined with thick, black plastic film. The bed must be watertight as leaks mean wasted nutrients. Polyethylene film (6-20micron), or swimming pool liner are ideal. If using black polyethylene film, it should be doubled over for maximum strength – a single layer of polyethylene will stretch and tend to mould around sharp objects once sand is shovelled into the bed, which may cause it to rupture.

The bottom of the trough should have a 1:400 incline (150mm drop per 60 metres), so that it can be drained or leached when required. A drain pipe should be set along the entire length of each sand bed, which in turn should be connected to a main pipe at one end to collect waste water from all beds and to conduct it away from the greenhouse or growing area.

For sand beds deeper than 400mm, 50mm agricultural pipe, covered with blue metal, can be used to channel excess nutrients away. For a standard sand bed, 70mm pressure pipe can be used and this should be seated on a shallow layer of blue metal. The pipe should have drainage holes, cut across the pipe with a saw every 450mm, with the holes positioned on the undersurface of the pipe to discourage plant roots entering the pipe. For greenhouse operations, the floor should have a gradient of 150mm per 30 metres, (1:200 incline), and should be covered with 6 micron black plastic film. Generally, two layers of black film are used to overlap and cover the entire floor.

A 30mm-50mm diameter drainage pipe, cut in the manner described earlier, is placed on top of the plastic at a uniform spacing of 1-2 metres between pipes, depending upon the grade of sand in use. The finer the sand, the closer the pipes should be spaced.

Drain pipes must run parallel with the slope of the floor. At the low end of the greenhouse a connecting pipe is installed to conduct excess nutrients away from the greenhouse where it can be used for outside irrigation. Once the pipes are in place, the entire greenhouse floor is covered with sand to a depth of 300mm. A simple home system can be designed along the same lines as commercial units, but on a much smaller scale. It should consist of a bed or growing tray, nutrient reservoir and a trickle feeding system operated by a pump that is controlled by a timer. The growing tray can have small holes in the bottom of the plastic liner or a perforated plastic pipe can be used for drainage.

One of the oldest and still popular hydroponic methods is to use a wick system. This consists of a double pot, one containing the sand and plant, and the other the nutrient solution. A fibrous wick is set into the growing pot about one-third of the way with the other end suspended in the nutrient solution. As the sand dries out, capillary action draws more solution through the wick to the plant root zone. Ideal Sand Aggregate Research shows that beach sand is usually too fine and causes puddling, indicated by water coming to the surface upon vibration of the sand.

Examples of puddling can often be seen in footsteps while walking along wet beach sand. It is caused by the high percentage of silt and fine sand. The ideal sand aggregate is river sand, washed free of fine silt and clay.

The sand particle size should be between 0.6mm and 2mm in diameter, which allows the aggregate to drain freely and not puddle after an application of water. Cuming recommends a mixture of grades (30-40 per cent of 0.5mm, 40-60 per cent of 0.2mm to 0.5mm, and 5-15 per cent of 0.2mm). Drip irrigation System A drip irrigation system must be used with sand culture. In a greenhouse situation piping should be capable of delivering 6-10 litres per minute for each 100 square metres, or 30-45 litres per minute for each 500 square metres of growing area. However, the rate and length of irrigation cycle will depend upon the crop, its maturity, weather conditions and time of day.

The volume of water can be regulated using a flow control valve for each sand bed. The valve is adjusted for each bed according to plant requirements. In this way several plant types can be grown. The flow valve should be positioned upstream from the solenoid valve which automatically controls the irrigation cycle. Where several beds are cultivated, there is no need to water them simultaneously. The water arrangement can be set so that individual beds are watered separately, thus ensuring mains pressure is not reduced. For small sand beds, a single 13mm black polyethylene pipe can be run down the length of the bed with spaghetti lines inserted every 300mm. Larger beds will require 13mm poly pipe run along the inside of each plant row.

Emitters can be used to deliver nutrients to plants, adjusted to deliver 4-6 litres per hour. Alternatively, spaghetti tubing can be inserted into the main 13mm feed line. A short length (50mm) of 13mm poly tube can be attached to the end of each spaghetti line to channel nutrients out to each side of the irrigation point. Emitters, above ground pipes and fittings should be black to prevent algae growth inside the piping system.

Watering

If a timer is used, it should be programmed to deliver nutrients two to five times daily, depending on the maturity of plants, weather and seasonal factors. The water is added to each cycle to allow 8 per cent-10 per cent drain off. Twice a week a sample of the drain off should be taken and tested for total dissolved salts. If the dissolved salts reach 2000 parts per million (ppm), then the entire sand bed should be leached free of salts using fresh water.

The Nutrient Tank

The nutrient tank should be large enough to meet feeding requirements for at least one week. And if several crops are grown which have different nutrient requirements, then two tanks should be used, each with its own specific nutrient formulation. In this event, independent irrigation systems must be connected to each nutrient tank. Since sand culture is an open system, there is no need to change the nutrient solution regularly. However, the tank should be drained and cleaned periodically of any sludge or sediment which may accumulate owing to inert carriers in the fertilizer salts.

Sand culture beds can be constructed on any surface, including hard bedrock or stony ground with a minimum of fuss. That, combined with low installation costs, simplicity of operation and retention of moisture, even in hot weather, makes it an ideal system for Australian conditions. It is particularly well-suited to rain shadow areas. Perhaps the only significant disadvantage is that it is likely to become waterlogged during heavy rain periods in those systems exposed to the elements.

ORGANIC GROWING MEDIA AND FERTILIZERS FOR GREENHOUSES

Organic methods of plant production are of increasing interest to many growers of horticulture crops including greenhouse operators. Some greenhouses are seeking "organic certification" in the same way as farmers of edible crops and livestock. The purpose of this chapter is to review the current status of "organic" growing media and fertilizers for greenhouse crop production. It is said that the true organic grower is seeking to produce crops in "balance" or "harmony" with their environment so that finding chemical pesticide and fertilizer substitutes will be unnecessary. I believe this might true in outdoor crop production, but in the rather unique greenhouse environment this might not be readily achieved. The greenhouse environment is very effective at promoting rapid plant growth, as well as insects and diseases in a short time period. Greenhouse crops are grown in small volume containers and they are irrigated frequently with large volumes of water. The combination of rapid plant growth, limited root volume, and frequent leaching make fertility management a challenge without the use of water-soluble chemical fertilizers.

Right now, because of the special challenges of growing in the greenhouse, it may be difficult for a traditional greenhouse grower to quickly transform into an "organic grower" without first looking for substitutes.

Personally I've got no problem with looking for organically acceptable substitutes for the traditional soilless growing media and chemical fertilizers. I believe that finding substitutes may be about the only way for most greenhouse growers have in immediate future to get into organic growing and that's what this piece is about. Probably the first thing that comes to mind when we talk "organic" is composting and using the compost as a substitute for traditional soilless media.

THE BASIC COMPOSTING PROCESS

The general steps in the biological process which creates compost are the same regardless of the raw materials being composted or the size and complexity of the production facility. A compost must pass through all of the steps outlined here in order for it to be considered of high enough quality for use in organic potting mixes. The progress of organic matter decomposition during composting can be followed by monitoring the temperature of the compost pile. During the initial phase of composting the temperature of the pile increases rapidly as the population and activity of decay microorganisms increases in response to the readily decomposable carbon in the raw materials. The goals are to reach a temperature between 130°F or more and to maintain this temperature range until the microorganisms begin to exhaust the readily available carbon. Heating to these temperatures is critical for high quality greenhouse compost because it is in this range that weed seeds and most plant and human pathogens that might come from animal manure are killed. During composting the pile is turned and remixed several times to ensure complete heating and decomposition.

Following the high temperature phase there is an extended period of gradual temperature decline until the pile reaches ambient air temperature. Now, if the pile is turned, reheating will not occur. At this point the compost is said to be "near maturity", but to ensure that the compost is stable and ready to use, most producers allow some extra time for the compost to "cure". How long composting lasts varies with the method. It could take about 1-2 years in a static unturned pile, 6-9 months if the pile is turned occasionally, or only 1-4 months the pile is turned frequently.

To comply with the National Organic Programme standards compost piles must maintain 131-170°F for at least 3 days (static pile) or at least 15 days (windrow, turned at least 5 times). High temperatures are necessary to kill any human pathogens especially if farm manure is a component. Also, weed seeds and plant diseases are most successfully killed at high temperatures.

Many types of raw materials can be used for making compost; some common materials are listed in the following table. Pay close attention to the comments in the table.

Table. Some Raw Materials for Making Compost

Material	Comments
Farm animal manure	Must be composted
	Straw and bedding
Crop residues	Must be pesticide free
Fruit and vegetable wastes	Must be pesticide free
	Food processing wastes
	Seafood processing waste
Grass clippings	Must be pesticide free
Sawdust and other wood wastes	Use in moderation, low nutrient value.
Newspaper	Black ink only, shredded, <25 per cent
Leaves	Shredded

It is important that the raw materials be properly prepared prior to mixing and the start of composting. Most organic materials must be shredded or ground to reduce particle size and help make them less resistant to decay.

During composting, oxygen and moisture levels are critical factors in determining the degree of decomposition which takes place and the length of time it takes to reach a stable product. Oxygen levels below 5 per cent and moisture levels above or below the range of 40-65 per cent inhibit the composting process. Most composting operations aerate the piles and irrigate them if conditions favour excessive drying. The volume of the finished compost is smaller than the volume of raw materials because of the breakdown of organic matter and the evaporation of water. When is the compost ready for use? Currently there is no single widely accepted criterion to determine when compost is "done". Measurements of temperature, respiration, ammonia production, pH, and carbon:nitrogen (C:N) ratio are among the potential indicators of compost maturity, but no one factor is completely reliable. Generally, at the end of active composting (heating period) producers allow a "curing" period of about 1-2 months to make sure the compost is stable before it's used.

GREENHOUSES WORK

Greenhouses are often seen as romantic structures. Originally the exclusive property of the wealthy and well-born, the first greenhouses were probably built in Roman times to cultivate exotic fruits and vegetables. In the first century, Pliny the Elder made a reference to the Emperor Tiberius having had a portable greenhouse that was protected with a covering made of transparent stone. This unusual and rare greenhouse was devised to cultivate the emperor's favourite vegetable, the cucumber.

Produce that we can find today in most local grocery stores was at one time considered priceless in many parts of the world. In the 17th century, entire buildings were erected to house and propagate oranges and pineapples. Before they were called greenhouses, the names for these structures were as exotic as the fruits they contained. They were called specularia, orangeries and pineries.

Reproducing plants out of season gave man a measure of control over nature. The allure of it sparked the imagination and inspired new methods for building structures devoted to plants. Precious glass began to be used more and more in greenhouse construction.

Harnessing the plant world and exploring the possibilities of cultivating useful, exotic species led to building larger and more elaborate greenhouses, some of which are still in existence today. This fascination with propagating plants in a controlled and protected environment has never dimmed. Greenhouses have grown from a novelty to an essential component in the way we feed hungry populations around the world, cultivate plants for medical research and preserve plants for future generations to enjoy. In this chapter, we'll see how greenhouses work to make all of this happen and why you may just want a greenhouse in your backyard.

USES FOR GREENHOUSES

A greenhouse can extend a plant's growing season by a few weeks, or it can create a complete microclimate that's a successful substitute for the plant's native environment. From its origins as an indulgence for the wealthy and privileged, greenhouses are now used in many important and unexpected ways to help man understand and make use of the natural world.

Since World War II, there's been a huge increase in the commercial use of greenhouses to feed the world's population. In temperate climates, greenhouses extend the growing season and protect plants from harsh weather conditions. In higher latitudes, greenhouses increase plant production by optimizing the available light. Even in hot, arid regions, specialized greenhouses have been developed that can help lower temperatures and manage water loss in plants due to transpiration.

Cultivating attractive and hardy plants for commercial nurseries and florists is a thriving industry that relies on greenhouses to create perfect, pest free plants.

In many areas, controlling the environment inside a greenhouse is easier and more reliable than trying to manipulate all the variables of growing delicate plants outdoors.

Universities and research facilities around the world, like the Eden Project in St. Austell, England, are using the controlled environments in greenhouses to recreate specialized ecosystems in order to understand plants better. Greenhouse technology is making it easier to study the potential value of medicinal plants and explore ways to increase plant yields and make plants more disease resistant. Greenhouses are even being used to preserve plant species whose natural habitat is threatened. So let's take a look at how they work.

HEATING A GREENHOUSE

Greenhouses create a sheltered environment for plants by using solar radiation to trap heat. This system of heating and circulating air helps to create an artificial environment in a greenhouse that can sustain plants when the outdoor temperature is too cool or variable. Heat enters the greenhouse through its covering of glass or plastic and starts to warm the objects, soil and plants inside. The warmed air near the soil begins to rise and is immediately replaced with cooler surrounding air that starts to heat up. This cycle raises the temperature inside the greenhouse more rapidly than the air outside, creating a sheltered, warmer microclimate.

In temperate climates, the sun might do all the heating in the greenhouse, but where the temperatures plummet, artificial heat may be necessary to maintain temperatures above freezing. Where some greenhouses have access to central heat from the main building, others have to rely on natural or bottled gas, heating coils or heating fans. These will usually work in conjunction with a thermostat. Because heat is one of the biggest expenses of keeping a greenhouse, other sources of energy are always being explored, like the use of solar batteries or animals as heat sources. There are also other processes acting on the air inside a greenhouse. The sun's energy can travel through greenhouse glass easily, but the radiation emitted by the plants and soil that have absorbed the heat doesn't get out as easily, helping to trap heat inside.

This makes it possible to keep a greenhouse warm, but it also can cause problems with overheating. In order to keep plants from getting too hot, some method of heat control is necessary. Vents that allow the lighter, hotter air to exit the greenhouse near the roof and cooler air to enter closer to ground level act as air conditioning. Proper ventilation keeps the air in a greenhouse circulating. This helps maintain a stable temperature and also cycles the carbon dioxide (CO_2) that plants need for photosynthesis. Generally, greenhouses have at least two vents, one on or near the roof and one on the lower half of the structure. Mechanical ventilators can also help maintain good airflow and heat control by opening and closing the vents automatically when the temperature in the greenhouse changes.

And of course, all of the plants inside a greenhouse need some sort of water. Whether you use a garden hose, watering can or a sophisticated automated system with water sensors, water is essential in a greenhouse. Because watering is the most time-consuming greenhouse chore, the use of some type of automated system, like wicking, capillary matting or drip irrigation can make the process more consistent and reliable. Even if feeding water directly to the greenhouse via an underground pipe isn't possible, placing a greenhouse near water is a practical necessity.

TYPES OF GREENHOUSES

There are many types of greenhouses, and this can include plastic greenhouses, rion greenhouses, those made of glass, temporary structures, permanent structures, and many others. The type you choose will depend on a number of factors. One important factor is the types of plants you intend to put in the building, because this will determine the temperatures required. If you are just looking for types of greenhouses to start your yard plants early in then a cold house or cold frame will work. These structures do not have any heating supplied at all, and they will only stay slightly above the temperature which is outside. These buildings can not be used all year long in colder areas where snow and freezing temperatures are common.

One of the most popular types of greenhouses are rion greenhouses, which are a form of plastic greenhouses. These are known for their incredible strength, and they can stand up to strong winds and heavy snows that would damage most other types of plastics. Portable or temporary types of greenhouses are usually very light, and they can be moved when the location is not the best possible for the plants being grown. These structures are usually very easy to put together and install, and they can be conveniently stored away when you are not using them. Greenhouses come in many materials, sizes, and types, and the ones that you choose should reflect your individual needs and requirements. This is a matter of personal preferences as well as the pros and cons, and each person may choose differently.

Since greenhouses use the sun to create a warm environment for plants, when the temperature drops, there's less sun-generated heat in the greenhouse. For plants that need more heat than a greenhouse can provide naturally, heating systems are necessary to make up the difference. Greenhouses are divided into categories based on how much supplemental heat they need to produce in order to keep plants at a certain temperature.

Cold Houses and Cold Frames: Cold houses provide protection for plants, but the temperatures inside can still dip below freezing during the winter because they have no supplemental heat source. Cold houses can help start spring crops a few weeks early and extend the growing season in fall, but they're limited by the weather.

Cool Houses: Warmer than cold houses, cool houses keep plants above freezing and in a temperature range of between 45 to 50 degrees Fahrenheit (7 to 10 degrees Celsius). Keeping the temperature above freezing will protect frost sensitive plants, like geraniums and hibiscus, which would be impossible to keep year round in areas that experience freezing temperatures.

Warm Houses: A warm house will allow a broader range of plants, but requires slightly warmer temperatures too, around 55 degrees Fahrenheit (13 degrees Celsius). Although the temperature range doesn't support many tropical plants, some varieties of orchids and ferns can over-winter in a warm house environment.

Hot Houses: These greenhouses are designed to house tropical plants like caladium, dieffenbachia and gardenia, which need a temperature range of 60 degrees Fahrenheit (15.5 degrees Celsius) and higher. They require the most supplemental heat and insulation and can be expensive to maintain.

Conservatories: Conservatories are designed to display plants, not just maintain and propagate them. They often have finished floors, ornate window treatments and space for furniture. Window greenhouses and small tabletop greenhouses are also considered conservatories because they're used primarily for display.

DIFFERENT TYPES OF GREENHOUSE

Greenhouses add growing time for plants in areas with extreme temperatures and weather conditions. The structures are built to house seedlings and mature plants for personal or commercial use. There are different types of greenhouse structures that either attach to a building or are freestanding units. The structures have a variety of shapes and covering materials from which to choose.

Attached Structures

A lean-to structure butts up against a building. Window mount and lean-to greenhouses are attached to a building, usually a house. The window mount is a small unit that is attached to the window frame in a building.

The lean-to structure has two types of roofing. The curved roof will allow the one-story greenhouse to be placed against a taller building. A straight-eave structure will fit up under the eaves and roofing system of a single-story house.

Even-span Greenhouse

The even-span greenhouse can be located at the end of an existing building. The open end of the full-size greenhouse is attached to the outer wall of the building. An entrance can be made into the building where the two structures connect. The even-span has a large area for growing and better air circulation than a lean-to.

Freestanding Structures

Freestanding greenhouse structures are built as self-supporting frames. They are made in styles such as Quonset, rigid frame, Gothic, A-frame, post and rafter. These different structures are built to stand on the ground or a foundation. Freestanding greenhouses are built in a variety of sizes.

Coverings

Greenhouses have a variety of coverings from which to choose. The home gardener may choose to build a greenhouse from recycled windows for a low-budget project. Glass has been traditionally used for years and can be a long-lasting cover.

Tempered glass is also an option for the builder. Clear or transparent, quality-grade fibreglass is a good option for a greenhouse covering as it is lightweight and can last 15 to 20 years. Plastic is used in double-wall and film forms. These are lightweight coverings with varying degrees of quality.

RESEARCH ON GREENHOUSE FOOD CROPS

In countries or regions where greenhouse production is an important industry, government and universities are generally involved in research on greenhouse production. The general objective of the research is to improve yield and quality of produce and profitability of production, by investigating all aspects of greenhouse production: greenhouse design and covering materials, growing methods, environment controls, substrates, plant nutrition, plant pathology, and insect control. Grower associations may also be involved in the development of research priorities, and may contribute financially to the expenses of research.

Due to the presence of a large and technologically advanced greenhouse industry in the Netherlands, the most notable research institutions are found there. The Research Station for Floriculture and Glasshouse Vegetables (under the Ministry of Agriculture, Nature Conservancy and Fisheries) has five sites. The other important Dutch institution is the University of Wageningen.

In the United Kingdom, Horticulture Research International (HRI), the largest horticultural research establishment in the world, maintains an active research programme on greenhouse crops and provides its services (from fundamental research to technology transfer) to research councils, government departments, growers, and commercial industries, in the European Community (EC) and other countries.

In the Americas, the Greenhouse and Processing Crops Research Centre (GPCRC; Agriculture and Agri-Food Canada) is the largest research facility specializing in greenhouse vegetables. The GPCRC is a leading member of the Canadian Network for Greenhouse Vegetable Research.

Japan, Spain, and Israel are some of the other countries with important research programmes in horticulture, including greenhouse food production.

The International Society for Horticultural Science (ISHS) is an international organization of horticultural scientists, which aims at promoting research in all branches of horticulture, including greenhouse food production.

Within the ISHS, there are various commissions and working groups related to greenhouse production.

FUTURE OF GREENHOUSE FOOD PRODUCTION

As the world population continues to increase, and more agricultural land is lost to urban development, intensive food production in greenhouses may play a more important role in food production. Furthermore, improving economic conditions in developing countries and an increasing preoccupation with health and nutrition will increase demand for high-quality food products. Through controlled climate and reduced pesticide use, greenhouses can meet this consumer demand. Foods with improved health characteristics or containing nutraceuticals (substances with pharmaceutical or health-beneficial properties that can be extracted or purified from plants) can be grown pesticide-free in greenhouses.

SELECTION OF COMMERCIAL GREENHOUSE FRAMES

A good selection of commercial greenhouse frames and framing materials is available. The frames are made of wood, galvanized steel, or aluminum. Build-it-yourself greenhouse plans are usually for structures with wood or metal pipe frames. Plastic pipe materials generally are inadequate to meet snow and wind load requirements. Frames can be covered with glass, rigid fibreglass, rigid double-wall plastics, or plastic film. All have advantages and disadvantages. Each of these materials should be considered—it pays to shop around for ideas.

FRAMES

Greenhouse frames range from simple to complex, depending on the imagination of the designer and engineering requirements. The following are several common frames.

- *Quonset:* The Quonset is a simple and efficient construction with an electrical conduit or galvanized steel pipe frame. The frame is circular and usually covered with plastic sheeting. Quonset sidewall height is low, which restricts storage space and headroom.
- *Gothic:* The gothic frame construction is similar to that of the Quonset but it has a gothic shape. Wooden arches may be used and joined at the ridge. The gothic shape allows more headroom at the sidewall than does the Quonset.
- *Rigid-frame:* The rigid-frame structure has vertical sidewalls and rafters for a clear-span construction. There are no columns or trusses to support the roof. Glued or nailed plywood gussets connect the sidewall supports to the rafters to make one rigid frame. The conventional gable roof and sidewalls allow maximum interior space and air circulation. A good foundation is required to support the lateral load on the sidewalls.

- *Post and rafter and A-frame:* The post and rafter is a simple construction of an embedded post and rafters, but it requires more wood or metal than some other designs. Strong sidewall posts and deep post embedment are required to withstand outward rafter forces and wind pressures. Like the rigid frame, the post and rafter design allows more space along the sidewalls and efficient air circulation. The A-frame is similar to the post and rafter construction except that a collar beam ties the upper parts of the rafters together.

Coverings

Greenhouse coverings include long-life glass, fibreglass, rigid double-wall plastics, and film plastics with 1- to 3-year lifespans. The type of frame and cover must be matched correctly.

Glass: Glass is the traditional covering. It has a pleasing appearance, is inexpensive to maintain, and has a high degree of permanency. An aluminum frame with a glass covering provides a maintenance-free, weather-tight structure that minimizes heat costs and retains humidity. Glass is available in many forms that would be suitable with almost any style or architecture. Tempered glass is frequently used because it is two or three times stronger than regular glass.

Small prefabricated glass greenhouses are available for do-it-yourself installation, but most should be built by the manufacturer because they can be difficult to construct.The disadvantages of glass are that it is easily broken, is initially expensive to build, and requires must better frame construction than fibreglass or plastic.

A good foundation is required, and the frames must be strong and must fit well together to support heavy, rigid glass. *Fibreglass:* Fibreglass is lightweight, strong, and practically hailproof.

A good grade of fibreglass should be used because poor grades discolour and reduce light penetration. Use only clear, transparent, or translucent grades for greenhouse construction. Tedlar-coated fibreglass lasts 15 to 20 years. The resin covering the glass fibres will eventually wear off, allowing dirt to be retained by exposed fibres. A new coat of resin is needed after 10 to 15 years. Light penetration is initially as good as glass but can drop off considerably over time with poor grades of fibreglass.

Double-wall plastic: Rigid double-layer plastic sheets of acrylic or polycarbonate are available to give long-life, heat-saving covers. These covers have two layers of rigid plastic separated by webs. The double-layer material retains more heat, so energy savings of 30 per cent are common. The acrylic is a long-life, non-yellowing material; the polycarbonate normally yellows faster, but usually is protected by a UV-inhibitor coating on the exposed surface. Both materials carry warranties for 10 years on their light transmission qualities. Both can be used on curved surfaces; the polycarbonate material can be curved the most. As a general rule, each layer reduces light by about 10 per cent. About 80 per cent of the light filters through double-layer plastic, compared with 90 per cent for glass.

Film plastic: Film-plastic coverings are available in several grades of quality and several different materials. Generally, these are replaced more frequently than other covers. Structural costs are very low because the frame can be lighter and plastic film is inexpensive. Light transmission of these film-plastic coverings is comparable to glass. The films are made of polyethylene (PE), polyvinyl chloride (PVC), copolymers, and other materials. A utility grade of PE that will last about a year is available at local hardware stores. Commercial greenhouse grade PE has ultraviolet inhibitors in it to protect against ultraviolet rays; it lasts 12 to 18 months.

Copolymers last 2 to 3 years. New additives have allowed the manufacture of film plastics that block and reflect radiated heat back into the greenhouse, as does glass which helps reduce heating costs. PVC or vinyl film costs two to five times as much as PE but lasts as long as five years. However, it is available only in sheets four to six feet wide. It attracts dust from the air, so it must be washed occasionally.

PRODUCTION SYSTEMS

Since the early days of greenhouses, plants have been grown in soil or in soil-filled containers. The first technique for fertilizing plants, which is still in use today in organic production, was the use of manure. Today, fertilization of plants can also be accomplished by incorporating chemical fertilizers in the soil, or by distributing fertilizers dissolved in water (so-called fertigation) to plants with a drip (trickle) irrigation system.

Intensive and repetitive cultivation of crops on the same soil generally results in a degradation of soil properties and fertility. Salt accumulation may be another problem in soil cultivation. Incorporation of manure, compost, and other organic materials into soil can be used to improve its structure and replenish its fertility. However, ensuring perfect fertilization of plants grown in soil is still a difficult task. Furthermore, intensive and repetitive cultivation of crops on the same soil can also result in insect or disease infestation. Soil replacement and soil fumigation are two solutions, but the first technique is expensive and the second is not always successful. Greenhouse production in soil is still used widely.

GROWING WITHOUT SOIL

In order to better control fertilization for optimizing plant growth and yield, and also to avoid the problems occurring in soil, growing systems that do not use soil (soilless) were developed for the cultivation of greenhouse crops. These soilless systems can be classified in two groups: liquid (water) and solid (artificial substrates that are either inorganic or organic). Systems using water as a growing medium are the nutrient film technique (NFT), deep flow technique (DFT), and aeroponics. Common inorganic media are rockwool, vermiculite, perlite, and clay pellets. Organic substrates are peat, coconut coir, sawdust, and straw. Inorganic and organic substrates are usually contained in bags, and plants are irrigated with a complete nutrient solution distributed by a drip irrigation system.

The excess of nutrient solution can either be allowed to leak into the ground or is recuperated and recirculated (after treatment) to plants. In liquid systems, plant roots are continuously exposed to nutrient solution, which is not leaked into the ground.

Growing methods using artificial substrates or water are known as soilless culture or hydroponics. Hydroponics is literally defined as the growing of plants in water, but the plants are actually grown in a complete nutrient solution. Ideally, the term *hydroponics* should be reserved for water culture, and the term *soilless culture* for plant cultivation on artificial substrates. In practice, the terms *hydroponics* and *soilless culture* are used indiscriminately to describe water and substrate-based systems.

Although official statistics are unavailable, hydroponic systems are known to be used extensively for food production in greenhouses. The most popular soilless medium for hydroponic vegetable production is rockwool. The nutrient film technique is also often used, but to a much lesser extend than rockwool. In some regions, the availability of low-cost materials may provide alternative substrates.

For example, in British Columbia, sawdust, a residue of the large forestry industry, is commonly used as a substrate. Both aeroponics and DFT remain in little use today.

INSECT AND DISEASE CONTROL IN GREENHOUSES

One objective of hydroponics is to avoid insects and diseases that may occur in soil. In a soilless culture system, such as rockwool, it is easy to remove infected plants.

However, spread of diseases can occur very quickly in systems where nutrient solution is recirculated. Methods such as filtration of the nutrient solution, and disinfection with ozone or ultraviolet light, have been developed to eliminate pathogens that may be present in the nutrient solution. However, these methods are often expensive and not completely effective.

Greenhouses are used to create and maintain an environment ideal for plants. However, this environment is often favourable for insects and pathogens too. In the past, the control of insects and diseases in greenhouses was accomplished with the use of pesticides, but over time both insects and diseases have developed resistance to such pesticides, while consumers have begun to demand pesticide-free produce. Biological agents are now used to control whitefly, thrips, aphids, and two-spotted spider mite in greenhouses; few reliable biological agents are currently available for the control of diseases.

GREENHOUSE HIGH PRESSURE HEATING SYSTEM

In this system it makes no difference whether the floor of your boiler-house is on the same level as the greenhouses or two feet below or two feet above. It will make no difference in the working of the system. In my case the ground at the north-east end of the range happened to be two feet lower than the average level of the greenhouse site. I mention this because I think the north-east is where a building such as a boiler-house necessitates will cast the least shade and do the least harm. Being almost on a level with the surrounding ground the fuel can be wheeled to the furnace doors, but better than that, the ashes can be wheeled out. There is no iron ladder to climb down and break your neck on Christmas eve or other occasions when a faltering step is excusable, no water or stopped up sewer to fret and annoy.

Now all there is about this system besides saving labour and expense in installation is that you carry from forty to sixty pounds of steam on the boiler all the time. One practical friend said he carried sixty-five pounds and believed it was more economical of fuel than forty pounds, and by the use of a controlling valve the pressure is reduced one, two, or five pounds on your system of heating. The controlling valve recommended to us was the Mason, made in Boston, Mass., and not once in three years' use has there been the slightest hitch or trouble with it.

The slightest turn of the brass key on top of the valve will increase or diminish the quantity of steam passing through the valve. A small steam clock connected with the main steam pipe a few feet beyond the controlling valve will tell you how much steam you have on your heating system.

In our case the main steam pipe leaving the boiler is only a 3-inch, carrying of course the fifty pounds of steam till it reaches the controller, which in moderately cold weather is set at, three, in severe weather at five or six, and this is enough to carry the steam in some of the houses for full 500 feet.

It is just worth mentioning here that while you have fifty pounds of steam in your boiler, if a sudden storm comes up or a sudden fall of temperature, all of which often happens, there is no great commotion at the fire door but simply a turn of that little brass key on the controller and the opening of a few valves. You have the reserve steam at all times and it costs no more to keep it there than it does to keep up two pounds of steam on your gravity system. Now, whatever arrangement of heating pipes you have there will be one pipe you call your main return, which in the gravity system empties the condensed steam into the boiler. In the pressure system it empties into an automatic steam pump or a trap. It is my duty to be honest, so I must confess I have no experience with a steam trap of any make, but I have watched them working and have consulted practical men who have used both the trap and automatic pump and in every case they gave a strong preference for the pump.

I understand that a trap does not work unless you have five or six pounds pressure on your heating system. In the case of the pump you can have but one on your radiating pipes. That will make no difference because your pump is supplied by a small pipe (1-inch will do) from top of your boiler - where it gets whatever pressure is in the boiler.

The return, pipe will most likely leave the greenhouse about the level of the floor and run as directly as convenient to the pump, where it falls into a drum or cylinder which in a small pump is about two feet long and fifteen inches in diameter. In this drum there is a float and as soon as there is a gallon of water returned the float rises and lets in a jet of steam, which starts the little pump working, and by a 1 1/4-inch pipe throws the water into the boiler smoothly and almost silently.

This pump keeps your system clear of any condensation. Our pump is in a small brick pit eighteen inches below the floor of the boiler-house. This can always be arranged, but it appears that is not necessary, and if the pump were set on the floor it would work perfectly, because the pressure behind the condensed steam would raise the water two feet into the drum. Wherever convenient, however, I would prefer to have the drum a few inches below the return pipe so that there would be decided fall into the drum.

In the chapter on gravity I said about all I could on the distribution of the radiating pipes, so what more I can say on the high pressure system will be devoted to remarks on its advantages. Although reduced in pressure by the controller the steam is still hotter and drier than steam only raised slightly above the boiling point.

When occasion arrives, it will travel to the farthest point of your system much quicker than by increasing the supply of steam with a stronger fire. In many places the draining of a deep pit is almost impossible without great expense. There is no need of it in the least.

It is money wasted. You will find that after this steam under the pressure system has been through your heating pipes it will return to the pump in the shape of steam, showing that you have got the utmost benefit from it, but the chief claim of the advocates of the high pressure system is that it is a decided saving of fuel, and I quite agree with them. Why this should be I am not able scientifically to argue or demonstrate, yet it seems that after you have once got the forty to sixty pounds on your boiler, very little additional fuel will hold it there. It is certain that any condensation returning to the boiler is much hotter and nearer steam than under the gravity system, hence less fuel is required to convert into steam. In conclusion, without an elaborate drawing it would be difficult to demonstrate the arrangements of pipes, position of the controller, pump, by-path and other points, but any one thinking of installing this up-to-date system would surely visit one of the many establishments where this system is in use. He could take quite a journey for the price of digging ten feet into the earth.

There is one point more. Whatever steam system you use, don't depend on one boiler. If you need eighty horsepower, buy two forties, if a hundred get two fifties, and it is a saving in fuel to have plenty of boiler power. As it is in hot water heaters, a little recent real experience may be worth quoting. For two years we heated about 22,000 feet of glass with one boiler rated at fifty horsepower. Then we built on another 12,000 feet of glass situated 250 feet from the boiler-house, and put in another boiler rated at seventy-five horsepower. Now on occasions either one of these boilers will heat and has heated the whole place, but both night and day firemen are positive that one-third less coal is burned when they are using both boilers to do the work.

What wonderful stuff this asbestos covering is and how it enables you to carry steam long distances in the open an! You have heard of the rival safe makers, one of whom, to demonstrate the imperviousness of his safe to heat, locked a mouse in the safe and then subjected it to great heat. On unlocking the safe the poor mouse was found frozen to death. A more truthful experience, and one nearer home, has astonished us; in running a 2-inch steam pipe from the boiler house to some carnation houses 250 feet distant, we have to cross an orchard ten feet above the ground.

The 2-ineh pipe is covered with asbestos an inch thick and then a thin covering of tar paper. We have noticed frequently that while five pounds of steam was coursing through the 2-inch pipe, icicles were hanging from the tar paper covering.

FOOD PRODUCED IN GREENHOUSES

Although greenhouses have been in existence since 1800 (or earlier), and greenhouse food production started to develop as an industry in the second half of the nineteenth century, the largest growth and expansion of the greenhouse industry occurred throughout the world following World War II. Today, food production in greenhouses can be found in all continents. Most popular food crops grown in greenhouses are tomato (beefsteak, cluster, Italian, cherry), cucumber, and sweet pepper. Other greenhouse grown vegetables include watermelon, muskmelon, summer squash, zucchini, lettuce, eggplant, snap beans, celery, cabbage, radish, Welsh onion, and asparagus. Fruits such as grapes, strawberry, banana, pineapple, papaya, orange, mandarin, cherry, and fig, as well as culinary and medicinal herbs, are also grown in greenhouses.

TODAY'S GREENHOUSES

The main greenhouse covering materials are glass and polyethylene (PE). Glass has been used since the early days of greenhouses. The introduction of

PE film after World War II was the main reason for the expansion of greenhouse production around the world, and it is now the most widely used covering material in the world. Glass-covered greenhouses are concentrated mainly in northern Europe and North America.

The low cost of the PE greenhouse is the main reason for its high popularity, especially in developing countries. In recent years, the use of PE-greenhouses has even spread to northern regions.

Research has shown that, under Canadian climatic conditions, heating costs of a double-layer PE-greenhouse are 20 to 30 per cent lower than for a glass-covered greenhouse. Most of the greenhouses built now in Canada are covered with PE. Standard PE film blocks the ultraviolet, but not the infrared radiation, and has a short durability.

However, improved PE films retain the infrared, but allow the ultraviolet, radiation (necessary for the bees, used for pollination of plants, to orient themselves) in the greenhouse, and are more durable. Polyvinyl chloride (PVC), another plastic film used to cover greenhouses, is used mostly in Japan. Other covering materials for greenhouses include rigid plastic acrylic, fibreglass, polycarbonate, and PVC panels, but their use is generally limited because of their high cost, compared to PE. Beside glass and PE, polycarbonate is often used on the sidewalls of polyethylene greenhouses in northern regions because of its good insulation, durability, and reasonable cost.

Greenhouses come in many styles and sizes, from the original houses with minimal climate control (furnace and vents) to the modern 10-ha (25-acre) or more, multispan greenhouses with high-tech climate controls (sophisticated and powerful heating system, CO_2 enrichment, evaporative cooling pads, exhaust fans, roof vents, thermal/shade curtain, computer controls, light sensors). Most sophisticated greenhouses are generally found in the developed, northern countries. Phytotrons are highly sophisticated structures that allow for accurate control of environmental conditions including light, and are generally used for scientific research in universities and research institutes.

However, phytotrons cannot be considered greenhouses since they are not covered with a transparent material. The degree of environment control needed depends on various factors. The first factor is the location of the greenhouse (local climatic conditions). Northern regions are characterized by cold winters and warm summers. If the objective is to grow plants all year long, then such large differences in climatic conditions between winter and summer require a high-tech greenhouse. In regions such as the Mediterranean (Spain, Italy, Morocco, Greece), the mild winter climate does not require the use of powerful heating systems, and low-tech greenhouses are sufficient for winter production. However, these regions have very hot summers, and the use of a low-tech greenhouse may not provide satisfactory temperature control to grow plants during summertime. The production schedule also affects the level of environment control and thus the level of technology. A greenhouse in northern regions may require a high level of climate control if the objective is to grow crops all year long (or long-season crops). If the objective is only to extend the production season (*e.g.*, one early crop in spring), then a less sophisticated greenhouse could be satisfactory.

Optimal growing conditions differ from one species to another. For example, lettuce prefers cooler temperatures than cucumber. Thus, the crop grown in the greenhouse may influence the level of environment control needed or desired. A low-tech greenhouse may provide sufficient climate control for lettuce but not for cucumber, depending on the location of the greenhouse and the production schedule.

Economic development also plays a role in the level of technology used in the greenhouse. In developing countries, growers may not be able to afford the most sophisticated equipment, and may lack technical expertise and technical support. Although greenhouses were developed in northern regions as a means of protecting crops against cold temperatures, and are therefore generally associated with cold climates, they are also used in arid regions such as Saudi Arabia. In such regions, the objective of the greenhouse is to protect plants from the excessive solar radiation and temperature, and to prevent excessive water loss by plants (especially since water resources are generally limited in those regions).

Therefore, technology in greenhouses in these regions is directed towards cooling. In northern countries, high-tech greenhouses can provide optimal growing conditions (temperature, humidity, carbon dioxide) for vegetable crops even during the coldest winter months. However, even with excellent climate control, yield and quality of crops grown during these months are low due to the low light level available. Research has shown that it is possible to produce good yield of high-quality produce during the winter months by using artificial light to supplement the natural radiation. The most common artificial lighting is the high-pressure sodium lamp. The high cost of electric energy in many regions is the most important factor preventing an increased use of artificial light.

GREENHOUSE HEATING BY HOT WATER

I must preface this very important subject by a confession. Some fifteen years ago I made the acquaintance of a steam-fitter, a clever workman, who had been quite successful in heating dwelling houses and thought he had ideas on greenhouse heating, but on that he was away off. A weakness of the writer is to be anxious to learn from any one who he thinks knows more on any subject than he does. So I learned from this man several fallacies on greenhouse heating which I have many times regretted putting into print. It does me good to confess this and I must ask the reader of the article on hot water in the first edition of this book to believe that where my present views conflict with what I wrote six or seven years ago, it is the result of later experience and practice. One of the fallacies was the use of small pipe, that is, less than 2-inch. Looking at this question superficially, the small pipe, 1-inch or 1 1/4-inch has the advantage over larger because you get more radiating surface in proportion to volume of water, but in practice this does not hold good.

Friction is so great that the circulation is retarded, radiation is rapid and the small amount of water in the small pipe soon cools. While 11/4-inch pipe may work very well in a house not over fifty feet, for any longer distance use nothing less than 2-inch.

The next and most costly mistake for me was running the pipe straight up from the boiler perhaps to within a foot of the greenhouse ridge, using a 2 1/2-inch or 3-ineh pipe, and at the farther end then dropping into a number of returns containing eight or ten times the volume of water of the flow pipe. Here is a dismal specimen of one of the houses, and I have reason to believe the same mistake has been made in hundreds of cases.

He had a 4-inch pipe rising from the boiler to the height of the greenhouse ridge, then branched off with a 3-inch and suspended the 3-inch about a foot from the ridge to the farther end. The house was used for soft-wooded pot plants, and was 20x100. At the farther end the 3-inch branched into two 2-inch pipes and dropped to the level of the 4-inch cast iron pipe under the benches, with which the house was formerly heated, four under each bench. By a complicated lot of fittings one of the 2-inch pipes was actually expected to heat the four 4-inch. Could anything be more absurd? Roughly estimated, the four 4-inch contained sixteen times as much water as the one 2-inch. Now this was an extremely absurd case, yet I have seen many others, where the same mistake exists differing only in degree. What were the principal features in this failure? The pipe near the roof at the boiler end of the house was very hot for perhaps twenty feet; then it gradually cooled and for more than half the length of the house there was not heat enough for any to be diffused among the plants on the benches. When the 2-inch pipe emptied into the four 4-inch returns, 16 slightly warmed them for a few feet, and then for seventy-five feet there was no-heat under the benches.

I must digress a moment here to say that although growers generally want no steam or hot water pipes beneath their benches of roses or carnations, they are of undoubted benefit to our soft-wooded plants like geraniums and begonias. Every one who grows bedding plants must have noticed this, but under the arrangement described we had little but cold, damp benches. There is a time of year, fall and spring, when this pipe near the ridge is useless. Days or nights when it may be 50 degrees outside you need a little fire heat to keep out dampness and keep up a healthy circulation of the atmosphere. While firing gently you will also want a little ventilation, and the heat from overhead pipe will entirely be wasted and your house left chill and damp.

There was an absurd controversy in the florists' papers some fifteen years ago on this subject of overhead heating. Some said that as the natural source of the heat was from above, meaning the sun, that our pipes should be near the roof, a great fallacy. Another piece of information which appeared some years ago in answer to a query was, " If you use hot water you need not excavate for heater, if you intend to use steam then you must keep the boiler well down below the level of the greenhouse." Nowadays we know that, broadly speaking, the very reverse of this is the truth, for it is the return, or cool, water pressing on the warmer and lighter water in the heater that is the power or cause of circulation, therefore the greater the height of the descending column of water the faster the circulation.

In truth there has been nothing new discovered of late about the circulation. There may have been in its application. Some forty-five years ago, perhaps before, there was published in London by Hood a volume on hot water. There has never been a better work on the same subject since. We may have found out better and cheaper modes of applying it than prevailed in his day, but all the laws of circulation which he demonstrates so finely are just the same today and always will be, for they are natural laws, and can never be altered. Hood says that the circulation of hot water was well known by the Romans, and used for heating their baths, so this wonderfully useful method of warming our houses did not originate in London, New York or Kalamazoo.

The cause of circulation is finely illustrated by Hood by having two columns of water, say two lengths of 4-inch pipe, each three feet high and connected at the bottom with a piece of pipe of the same size. Half-way in this connecting pipe there is a valve. Fill one pipe with water at the temperature of 45 degrees, the other with hot water at a temperature of 180 degrees. Then open the valve connecting the hot and cold columns. The hot water will rise or overflow. Now this is all there is to the circulation of water in your system. The water in the cold column or your return pipe is denser and of greater specific gravity, in simpler words, heavier and forces the lighter and warmer water out through the flow.

Now this illustrates the motive power that first starts the circulation of hot water. It is the difference between the weight of the water in the return pipe and that in the boiler. The water in the boiler being made lighter by the fire, the colder and heavier water forces it up and it is replaced with cold water, so it must follow that the higher, and consequently heavier, the column of water in the return pipe the faster will be your circulation. And it follows again that the faster the circulation the hotter will your pipes be, for the water returning quickly to the fire has not time to get cool. "When your return pipe near the boiler is nearly as hot as the flow where it leaves the boiler your circulation is perfect. All of which goes to prove that the lower the boiler the better the apparatus will work.

Reserve all your drop till you get near the boiler and then drop perpendicularly down. This talk about giving the pipes a rise of one foot in a hundred, or the same drop, is all bosh. If the pipes were a dead level in the house it Mould be perfect, but it is better to havè a rise or fall of two inches in a hundred feet because you want when emptying the pipes to have a drain out. Providing your boiler is well down, and that is the very essence of the whole job, it makes no difference whether you have a slight rise in the flow pipes in the house or a slight fall.

As admitted at the start of this chapter, I was the victim of false theories and it took a few years to break away and return to principles I knew of many years ago. Now here is what we did some fourteen years ago and it was wrong. The top of the heater was some two feet six inches below the joists of the boiler shed. It would have been lower could drainage have been secured. As it was we had to lay a 4-inch socket tile 650 feet. The two 3-inch flow pipes rose straight up through the floor and eight feet higher. This was done to avoid the doorway leading from the shed into the greenhouses. Please imagine these two houses to be 20x125, running east and west and connected by wooden partitions and wooden gutter. The flow pipe in the houses was a 2-inch on each side of the houses, running a foot below the gutter and on the sides, on the posts a foot below the plate from which the bars spring; that is a good place for them, but it is by no means overhead heating, it is resisting the cold at a very important spot.

As we had carried the flow pipe from the boiler so high in the shed, we had to drop again to the level of the flow pipes in the houses, which was done with two elbows and a short piece of 2-inch pipe. Although it worked, this was against correct principles. The hot water had no natural tendency to descend, it would rather flow out on a level. At the farther end of the 125 feet the 2-inch flow went into a manifold with five 1 1/4-inch openings and they were the returns.

This was not anything like as wrong as the case described earlier, yet there was not hot water enough in the 2-inch to supply the five 114-inch. That and other causes, friction and rapid radiation, all combined made it a poor job. There was no need of air pipes or petcocks in this system because a 3/4-inch pipe was tapped into the highest point of pipes in the shed and led up to a large tank which supplied us with water for the houses.

This tank held 150 barrels of water and from it a 1 1/4-inch pipe was connected with the return pipe close to the boiler. It may have given us a constant pressure of fifteen pounds. Please don't attach the slightest importance to the fact that this feed pipe was from a vessel holding 150 barrels of water. Pressure in a liquid depends entirely on height, not bulk. A funnel thirty feet high and three feet in diameter at the top, tapering to one inch at bottom, would give no more pressure at the one inch opening than a column of water one inch in diameter and the same height as the spreading funnel. Now we ran along several years with the arrangement above described, thought it perfection, and read that others in the east had adopted the same principle. Yet now we feel sure that many tons of coal were wasted in trying to get those return pipes hot. With the small amount of radiation we wanted them hot, not lukewarm.

So two years ago last summer we made a great change, not with the slightest feeling of doubt or uncertainty, because we were only going on the sound laws and principles of hot water circulation. Needed a new heater, wrongly called boiler, and purchased the Burnham. Let me say here that I have reason to deeply regret recommending boilers in the past without sufficient experience with them.

Still, after two years' trial I must say that the Burnham boiler has proved most satisfactory. It is of heavy castings, simply put together and easily and thoroughly cleaned. In place of one 2-inch flow pipe we hung two 2-inch and at the farther end ran each flow pipe into a manifold with three 1 1/4-inch openings. The 2-inch could amply supply the three 11/4-inch and they were hot back to the shed. I should have started at the boiler.

Will say now that the Burnham boiler, or the size we have, has a 5-inch opening for flow, same for return. We ran up as near as possible to the joist with a 5-inch pipe, then branched with a tee into a horizontal 4-inch pipe. I now remember there were eight 2-inch pipes to be supplied from this 4-inch. On reaching the south side of the north house we branched with a tee into a 3-inch and rose to the level of the flow pipes. Remember, the 4x3x3 tee came out of a 4-inch horizontal, not an upright, pipe. After supplying that run the flow was reduced to 3-inch and went on to supply the north run of the north house. From the shed end to the farther end the 2-inch flow pipes have a rise of two inches. On reaching the manifolds provision had to be made to let air escape, so we tapped into each manifold a small petcock.

For the first ten days after we begin to fire with a new supply of water there is a considerable accumulation of air. After a week or two there seems no more in the water, and if these petcocks are neglected for weeks no harm results. This I have found repeatedly to be the case. You may not have a tank of water raised up thirty feet, and there is no need of it. A barrel or box holding twenty gallons of water, or a metal boiler, such as is used in our kitchens, will do equany as well. Let it be seven or eight feet above the bottom of your heater, and when there is a gallon of water in the barrel your system is full. In this case you do not need petcocks, but in place of them simply tap in a 1/2-inch pipe and let it run up a rafter or anywhere out of the way; that, of course, will constantly and with-, out watching vent the heating pipes of any air. See that the top of this little pipe is a foot or two above the highest point of the water in your feeding cistern. This rearrangement of pipes was a great satisfaction and unqualified success, so much so that in seven or eight other houses where the overhead and rise and fall delusion had existed it was all pulled down and common sense and a smooth, quick circulation exists.

If an amateur came to me and asked for advice about heating a house 20x100, or 20x75. and wanted it so arranged that he could leave it at 11 p.m. till 6 a.m., I should without hesitation tell him to get plenty of boiler power and use the 4-inch pipe for radiation. Use the heavy cast iron pipe used by horticultural builders, 9-foot lengths and four inches outside diameter, not what many greenhouse men have used, Mott's soil pipe, 5-foot lengths, four inches inside diameter; this is a thin, brittle metal, unsatisfactory in many ways.

I have had no experience with water under pressure, unless the pressure of our city mains constitutes that system. Our city water has a pressure of about thirty-five pounds to the square inch, and in my experience it is a very cheap heating system for a store or office. Wishing to heat a flower store in this city, which is something like 19x80 feet in area, I put a small heater in the cellar. It is simply three lengths of 3-inch pipe, each about three feet long, and run into a manifold at both ends. The coil is resting on two 4-inch brick walls about two feet from the floor and is bricked over top, sides and ends. One end of this coil is raised about three inches and from it rising to the floor is the 11/ 4-inch flow, which leads off, and by the help of some tees connects with three radiators on the floor of the store, and from the other end of the radiators the returns drop to the lower end of the coil. There are two natural gas burners under this very simple heater, which in the coldest weather has never been turned on more than one-third its force.

A 1-inch pipe from the city water is connected with the lowest part of the coil and the valve is never closed, so there is always a pressure of thirty-five pounds on the pipe and radiators. The highest part of the system is the top of the radiators, and in them is a petcock which should be opened every day to let out air, but often is not for weeks, and in a radiator it is not of so much consequence. There is nothing more about it, only the radiators can be made very hot; a great success. If a strong fire should expand the water in the heater it has to find room by driving the water back into the mains. The whole thing cost less than $50, and $5 worth of gas was consumed in the coldest month. Now this system could be used with great success wherever you have a boiler that would stand the pressure. You could use it on either the uphill or downhill systems, but you could not have any open air vents, and unless you trusted to the automatic air valves you would have to daily open the petcocks at the highest point. The architecture of many greenhouse establishments, the smaller and older places more especially, is so complicated and houses are on so many different levels, that sound advice as to how to arrange pipes is impossible unless you are on the spot. Three rules can be accepted as sound. Provide plenty of capacity in both flow and return pipes. Don't use less than 2-inch pipes, and reserve all the drop on your return pipes till just before you enter the boiler. Watch these points and you will save coal.

On a cold, wet, windy, winter's day there is nothing more therapeutic than pottering about in a greenhouse. A cup of tea, packet of biscuits, the radio and some seeds to sow or cuttings to take; life doesn't get much better. The trouble is such simple horticultural pleasures don't come cheap. Depending on the temperature which has to be maintained and the severity of the weather greenhouse heating bills can quite literally go through the poorly insulated glass roof and sides. Due in part to the high cost of heating many greenhouse gardeners only turn their heating systems on towards the end of winter or early spring. For those without any form of heating now is the time to decide which type of fuel to use and the best kind of heater to suit their plants and budget.

9

Forestry Horticulture System

ANTITUMOR COMPOUNDS

The plant kingdom is one of the attractive sources of novel antitumor compounds. The National Cancer Institute in the U.S., for example, has conducted an intensive screening programme since 1955 and has identified various potent compounds from higher plants. These antitumor compounds include maytansine, tripdiolide, homoharringtonine, bruceantin, ellipticine, thalicarpine, indicine-N-oxide, and baccharin. In addition to these compounds, some of the higher plant products such as vinblastine, vincristine, podophyllotoxin derivatives including etoposide, and camptothecin and its derivative have already been marketed as very important anticancer drugs. Taxol from Taxus brevifolia and related plants, is one of the most exiting compounds and was marketed in 1992. However, the concentrations of these active compounds in plants are generally low, the growth rate of the plants is slow and the accumulation pattern of these compounds is highly susceptible to geographical or environmental conditions. Therefore, it is not an easy task to produce economically these compounds by extraction from intact plants.

Homoharringtonine

Homoharringtonine together with harringtonine and isoharringtonine, were isolated from Cephalotaxus harringtonia by Powell *et al.* These alkaloids are complex esters of the inactive alcohol, cephalotaxine. These compounds inhibit the growth of murine leukemias, L1210 and P388, and KB cells. Homoharringtonine also shows activity against colon tumors, melanoma and leukemia in mice.

Since the chemical synthesis of homoharringtonine was not efficient for commercialization, studies of tissue culture were conducted by Delfel and his colleagues. They detected about 5-10 mg of the alkaloids per kg dry wt. of the callus tissues cultivated for 3 to 6 months. The levels were approximately 1 to 3 per cent of the concentrations found in the parent plant. These products were cephalotaxin, homoharri-ngtonine, harringtonine and isoharrin-gtonine. MS medium containing 1 mg/L kinetin and 3 mg/L NAA was favourable to induce the callus from C. harringtonia, while the medium without growth regulators strongly promoted organogenesis. The radioimmunoassay established by the same group showed the levels of cephalotaxine and its esters were only 1/300 of the intact plant.

Taxol

Under the intensive NCl screening programme of antitumor compounds in the U.S., Wall *et al.* began to isolate an active principle against KB cells from a tree, Taxus brevifolia in 1965. In 1969 pure taxol was first isolated and its chemical structure was disclosed in 1971. It is a diterpene amide and has shown against B16 mouse melanoma tumor, the MX-1 human mammary xenograft and CX-1 colon xenografts. The mode of action of taxol is rather unique because it stabilizes microtubles and inhibits depolymerization. The clinical trials begun in 1983 have shown positive results in the treatment of advanced ovarian cancer and breast cancer as well.

The FDA in the U.S., has approved taxol (generically known as paclitaxel) at the end of 1992; Bristol-Myers Squibb produces the drug for use in the treatment of ovarian cancer in patients who have failed to respond to other chemotherapies. Taxol is now being manufactured by extraction from the bark of wild-grown T. brevifolia trees. The demand for taxol will be undoubtedly increasing since it will be applied for other cancers including breast and lung cancers in the near future, but its supply is limited. Other related plants such as T. canadensis and T. cuspidata also contain taxol and other related compounds. Generally the concentration of taxol in Taxus plants are very low. Therefore, harvesting Taxus trees for production of taxol commercially causes a serious problem in the U.S., from the environmental point of view.

As alternative ways, plant tissue and cell cultures as well as chemical transformation processes from baccatin extracted from needles of the plant and the total chemical syntheses have been investigated in many research groups. In the U.S., Phyton Catalytic Inc. and ESCAgenetics announced a couple of years ago that they were establishing plant cell culture processes to manufacture taxol. The latter showed a photograph of a vial containing taxol powder produced by plant cell cultures. However, both companies have never published details of the plant cell culture procedure in scientific periodicals or at any meetings, although ESCAgenetics described some of their results using a couple. A couple of patents have been published. For example, a U.S., Patent filed by A.C. Christen *et al.* of U.S.D.A. described that the tissue of T. brevifolia had been successfully cultured to produce taxol, related alkaloids and alkaloid precursors, but the patent claims are exceedingly broad and identity of what the actual process was remains difficult to ascertain. Shuler *et al.* at Cornell University showed that a cell line of T. brevifolia in suspension provided by USDA-

Agriculture Research Station and Phyton Catalytic, produced 3.9 mg/L taxol in the medium after 26 days of culture. The level of taxol was determined by reverse-phase HPLC. They found that fresh cell weight increased by 4-fold after a lag phase of about 4 days and that taxol was first seen at 13 days, and increased sharply after 20 days. It is of interest that all taxol produced was secreted into the medium, which was very unusual for plant cell cultures. Wickremesinhe and Artheca of the Pennsylvania State University established callus cultures and suspension cultures from T. brevifolia cv Repandens, T. cuspidata, T. media cvs. Hicksii and Densiformis. Though some cell lines grew fast and their doubling times were 9-14 days, the levels of taxol were too low for commercial production. Induction of the hairy roots of Taxus plants has been tried.

DiCosmo and his colleagues of the University of Toronto in collaboration of Nippon Oil Co. Ltd. in Japan described that they detected and identified taxol in the callus of T. cuspidata and its level was 0.02+0.005 per cent in dry weight after 2 months in culture. Suspension cultures of T. cuspidata were also established from the callus cultures and subsequently immobilized onto glass fibre mats. The cells were maintained as immobilized cultures for 6 months. The level of taxol in the immobilized cells was 0.012+0.007 per cent of the extracted dry weight.

Taxol is one of the most suitable and desirable plant products to plant cell culture research because the shortage of its supply and its high-value. Bristol-Myers Squibb is therefore, financially supporting many research groups to establish the process of cell culture production of taxol.

The root of Panax ginseng C.A. Meyer, a perennial herb, so-called "ginseng" has been widely used as a tonic and precious medicine since ancient times particularly in oriental countries including Korea and China. It is effective for gastroenteric disorders, diabetes and weak circulation, and has been used as an adjuvant to prevent various disorders, rather than a medicine to cure disorders. Thus, ginseng has been recognized as a miraculous medicine in preserving health and longevity. It has been known that the root contains various saponins and sapogenins. Among them, ginsenoside-Rb has a sedative activity, while Rg has stimulative activities.

Although there are several species of ginseng, the commercially important species, P. ginseng grows in an area of 30-48? north latitude such as Korea. The cultivation of ginseng in the field requires four to seven years, and it is impossible to plant consecutively for 20 to 50 years but the demand for the plant has increased dramatically in the world, and its price has soared. These are reasons why many researchers have tried to produce ginseng cells through plant tissue cell cultures.

Furuya *et al.* at Kitasato University have studied P. ginseng callus tissues since the early 1970's. Meiji Seika Kaisha in Japan investigated the large-scale production of the cells established by Furuya using various types of fermentors. According to their patent published in 1973, crown gall calli, callus tissues and redifferentiated roots of P. ginseng were able to accumulate saponins and sapogenins known from the intact plant. The callus tissues and roots were cultivated on both MS solid and liquid media containing vitamins, sucrose, 2,4-D and suitable natural nutrients such as soybean powder or beef extract for several weeks at 25-28 C. The concentrations of crude saponins in the callus (21.1 per cent), in the crown gall (19.3 per cent) and in the redifferentiated root (27.4 per cent) were much higher than those in the natural root (4.1 per cent). The saponins were found to contain ginsenoside-Rb and -Rg. It is obtained cultured cells of P. ginseng containing ginsenosides. Choi of Korea Ginseng and Tobacco Research Institute has investigated *in vitro* culture of P. ginseng extensively. He indicated that plant growth regulators such as 2,4-D and kinetin in the medium affected the levels of saponins in callus and suspension cultured cells. For example, 3.62 per cent of total saponins was detected in the callus cultivated in MS medium containing 5 mg/L 2,4-D and 1 mg/L kinetin, while 8.78 per cent was produced in 10 mg/L 2,4-D and 1 mg/L kinetin medium.

After the Meiji Seika abandoned development of the ginseng project, another Japanese company, Nitto Denko Corporation, constructed a 20 KL fermentor to scale-up cultivation of ginseng cells in collaboration with Furuya in the middle of the 1980's. The company has optimized various environmental conditions using 30 L jar fermentors for the cell line having partly differentiated tissues originally induced. Glucose in the medium promoted cell growth in the initial stages of the fermentation and sucrose fed during the growth cycle stimulated the productivity of saponins. Although a higher NO2/NH3 ratio was favourable to the growth, it decreased the concentration of saponins in the cells. The growth was suppressed by moderate agitation, but the yield of saponins increased. The highest cell mass, 19 g/L on a dry weight basis was obtained using a 2 KL fermentor and the production rate of the cell mass was approximately 700 mg/L/day.

The chemical variation between mother plants of P. ginseng and its tissue cultured cells. Ushiyama indicated that their cultured cells contained basically the same constituents as those in the intact plant. Acute virulence tests, Ames tests and dietary tests with livestock feed containing 12.5 per cent of dried cultured cells for 6 months did not show any abnormalities in animals.

Camptothecin

Camptotheca acuminata, a native of North China, was found to produce a potent antitumor alkaloid, camptothecin, by Wall *et al.* It is highly active in Walker 256 rat carcinosarcoma and mouse leukemia. The clinical trials in patients with gastrointestinal cancer were at first very promising but subsequent trials showed toxicity.

Sakato and Misawa (184) induced callus from the stem of C. acuminata on MS solid medium containing 0.2 mg/L 2,4-D and 1 mg/L kinetin. The callus was transferred to the liquid medium. Gibberelin, L-tryptophan and conditioned medium stimulated growth of the cells. After 15 days of cultivation in suspension, the con-centration of camptothecin in the cells was 0.0025 per cent on a dry weight basis, which was about 1/20 of the level in the intact plant. A.J. van Hengel *et al.* in the Netherlands established the suspension culture system of C. acuminata and detected camptothecin in the cultured cells using TLC, HPLC and GC-MS. The highest level, 0.998 mg of camptothecin per litre of the medium, was accumulated in the cells cultivated in MS medium containing 4 mg/L NAA.

Podophyllotoxin

Podophyllum pelatatum, May apple, which is a common herb in eastern North America contains an antitumor lignan, podophyllotoxin. It is active to KB cells and is used against certain virus diseases and skin cancer. A semi-synthetic derivative of podophyllotoxin, etoposide (V-16), was found to be active against brain tumor, lymphosarcoma and Hodgkins' disease and was approved by the FDA in the U.S.

Bristol-Myers Squibb is one of the largest manufacturers of the drug. Production of podophyllotoxin by P. pelatum cell cultures was first attempted by Kadkade and he found that a combination of 2,4-D and kinetin in the medium supported the highest amount of its production. Red light stimulated the production. The Nippon Oil induced embryogenic roots from a callus of the plant in a liquid MS medium supplemented 1 mg/L NAA, 0.2 mg/L kinetin and 500 mg/L casein hydrolysates. The roots were then transferred to the medium without growth regulators. They detected 1.6 per cent of podophyllotoxin in the dried tissues, which was 6 times higher level than that in a mother plant. To increase the yield of podophyllotoxin, Woerdenberg *et al.* in the Netherlands added a complex of a precursor, coniferyl alcohol, and ß-cyclodextrin to Podophyllum hexandrum cell suspension cultures.

Addition of 3 mM coniferyl alcohol complex gave 0.013 per cent podophyllotoxin of the cells on a dry weight basis but the cultures without the precursor produced only 0.003 per cent. β-D-glucoside of coniferyl alcohol, coniferin, was a more potent precursor in terms of the yield of the anticancer compound (0.055 per cent), but unfortunately this compound is not commercially available. The same authors reported that cell suspension cultures of Callitris drummondii (conifer) also accumulated podophyllotoxin-β-D-glucose. In the dark, the cells produced approximately 0.02 per cent podophyllotoxin of the dry cell mass and 85-90 per cent of the lignans were the β-D-glucoside form, while in the light the yield of podophyllotoxin-β-D-glucose increased to 0.11 per cent. It is reported that callus tissues and suspension culture cells of Lilium album produced podophyllotoxin. One of the cell lines produced 0.3 per cent podophyllotoxin of dried cells together with small amounts of 5-methylpodophyllotoxin, lariciresinol and pinoresinol after 3 weeks of cultivation. The callus tissue induced from P. hexandrum was reported by Heyenga to produce podophyllotoxin, 4'-demethyl-podophyllotoxin and podophyllotoxin-4-0-glucoside when the callus was incubated in B5 medium containing 2,4-D, gibberellic acid and 6-benzylaminopurine. The levels of podophyllotoxin and its derivatives were similar to those in the mother plant.

PHARMACEUTICALS

ALKALOIDS

A variety of alkaloids have been used as pharmaceuticals and most of them are plant metabolites. The typical tropane alkaloids, atropine, hyoscyamine, scopolamine and cocaine, were widely used as blockers of the parasympathetic nervous system such as anodyne and antispasmodic. Research on production of these useful alkaloids by plant cell cultures has been carried out for more than 25 years, however, industrial production has not yet succeeded because of low producing ability of the cultured cells. Plants used for these studies are mainly Atropa belladonna, Hyoscyamus niger, Datura meteloids and others.

MORPHINAN ALKALOIDS

Codeine is an analgesic and cough-suppressing drug and Papaver somniferum L.is a traditional commercial source of codeine, and morphine which can be converted to codeine. Mature capsules of P. bracteatum accumulates up to 3.5 per cent of thebaine which also can be converted to codeine. Although many researchers have tried to produce codeine by undifferentiated cells of these plants, little success has been achieved.

It is indicated that morphogenetic differentiation from cultured cells of P. bracteatum was prerequisite for producing higher levels of thebaine. The researchers also showed the similar results as Kamimura's report using P. bracteatum and P. somniferum. For example, it is reported that cells of P. somniferum produced norsanguinarine, sanguinarine, cryptopine and other various alkaloids, but not codeine and morphine.

Concerning sanguinarine, shows that fungal mycelium of Botrytis sp. elicited production of sanguinarine by P. somniferum cells. The level of this alkaloid increased 26 times in the presence of the elicitor, 29 per cent of the dry cell weight, compared to the medium without it. Since the alkaloid extracted from intact plants is added to toothpastes as an anti-plaque agent, the commercial production of sanguinarine by cell culture technology was intensively investigated by Kurz's group at the National Research Council of Canada and a company in the U.S., several years ago, but the process has not yet been commercialized. The efficient production of thebaine and codeine using cell culture systems by de novo synthesis was not successful, therefore studied the biotransformation of codeinone to codeine using immobilized cells of P. somniferum. The conversion yield was 70.4 per cent and about 88 per cent of codeine converted was excreted into the medium.

BERBERINE

Berberine is an isoquinoline alkaloid which is distributed in roots of Coptis japonica and cortex of Phellondendron amurense. Berberine chloride is used for intestinal disorders in the Orient and it takes 5 to 6 years to produce Coptis roots as the raw material.

Mitsui Petrochemical has improved the productivity, and found that addition of 10^{-8} M gibberellic acid into the medium stimulated berberine productivity up to 1.66 g per L of the medium. Using a cell sorter and protoplasts of C. japonica, they selected many higher alkaloid-producing cell lines. Thus, the Mitsui group produces berberine in a large scale at a level of 1.4 g per L of their optimized medium within 2 weeks. Furthermore, they established a "high-density cell culture" process to produce berberine much more efficiently. In order to achieve a cell mass of 70 g/L on a dry weight basis, stirring without damaging the cells, supply of sufficient amounts of oxygen, and that of appropriate nutrients were optimize. As a result, the yields reached 70 g per L of cell mass and 0.45 g/day of berberine. The continuous culture with high cell density was also conducted successfully.

Addition of a polyamine, spermidine, was found to stimulate the production of berberine by Thalictrum minus cell suspension cultures by Hara *et al.* in Tabata's laboratory although other polyamines such as cadaverine, putrescine and spermine were ineffective. They indicated that spermidine effected an increase of ethylene generation which was associated with activation of berberine synthesis. The maximum stimulative effect was obtained by addition of 2 mM spermidine.

TROPANE ALKALOIDS

Scopolamine and hyoscyamine are being used commercially as anesthetic and antispasmodic drugs. These alkaloids occur in leaves of (Solanaceae) plants including D. myoporoides and D. leichhardtii. Scopolia, Atropa, Hyoscyamus and Datura also contain tropane alkaloids.

Studies on production of tropane alkaloids by plant tissue cultures have been actively carried out by many researchers since West *et al.* found tropane alkaloids in an Atropa belladonna callus more than 30 years ago. However, the concentrations of scopolamine and hyoscyamine in cultured cells are generally very low in spite of many efforts to increase the yield using various approaches. Therefore, the plant cell culture has not yet been employed to manufacture these alkaloids. For example, Tabata *et al.* added tropic acid into Scopolia japonica suspension cultures as a precursor and could increase the level of alkaloids up to 15 times. Mitsuno *et al.* selected a high tropane alkaloid producing strain of Hyoscyamus niger which produced about 7 times more hyoscyamine (13.9×10^{-3} per cent fresh cells), than that of the parent strain. According to their results, there was no direct correlation between high producing ability and variation of chromosomal numbers.

L-DOPA

L-DOPA, L-3,4-dihydroxyphenylalanine, is an important intermediate of secondary metabolism in higher plants and is known as an precursor of alkaloids, betalain, melanine, and others. It is also a precursor of cathecolamines in animals and is being used as a potent drug for Parkinson's disease.

It is found that the callus tissue of Mucana pruriens accumulated 25 mg/L DOPA in the medium containing very high concentration of 2,4-D. Wichers *et al.* immobilized cells of M. pruriens within alginate and found that the cells produced DOPA from tyrosine in up to 2 per cent of dry cell weight. The DOPA synthesized was secreted mostly into the medium.

Teramoto and Komamine induced callus tissues of Stizolobium hassjoo (Mucuna hassjoo), M. pruriens and M. deeringiana, and optimized the culture conditions. The highest concentration of DOPA was obtained when S. hassjoo cells were cultivated in MS medium containing 0.025 mg/l 2,4-D and 10 mg/l kinetin. The level of DOPA in the cells was about 80 nmol/g-f.w.

Cardinolides

Cardiac glycosides or cardenolides are products of Digitalis species. Some of these compounds have been employed in treatment of heart diseases. Chemical structures of the major aglycone of cardenolides and various sugar residues link the 3-hydroxy group of aglycones.

For commercial production Digitalis plants are being cultivated in the fields of several countries including the Netherlands, Hungary and Argentina. Although there are a number of papers describing production of cardiac glycosides in Digitalis tissue cultures, generally the yield was very low, and moreover, during the successive transfers of the cultured cells the amount of cardenolides often decreased and disappeared completely.

Instead, many researchers indicated that morphological differentiation caused an increase in productivity. For example, it shows that organ cultures of D. lanata leaves and roots produced cardenolides and during the cultivation the level of digoxin in the tissues rose with increasing age. Hirotani and Furuya also found that renewed organ differentiation from callus tissues of D. purpurea led to a new formation of cardinolides.

Nutritional factors including growth regulators, sugars, nitrogen sources, vitamins and so on in the medium affect on differentiation of shoots and other organs and the secondary metabolites such as cardiac glycosides are often synthesized. The differentiated tissue culture is prerequisite for secondary metabolites production in some cases, however in general, the method requires much longer culture time than the suspension cell culture and consequently it is not efficient. Various secondary metabolites other than digitoxin and digoxin have been found in callus tissues of D. lanata and D. purpurea. These include cholesterol, campesterol, stigmasterol, β-sitosterol, 4-hydroxy-digitolutein and others. It is found that the levels of those secondary products tended to decrease over several passages in cultivation. In contrast to the de novo synthesis, the biotransformation process with Digitalis plant cells seems to be more promising from a commercial point of view.

It is reported that D. lanata and D. purpurea callus cultures rapidly transformed progesterone to pregnane. Leaf and root cultures of D. lanata and shoot-forming callus tissues of D. purpurea accumulated an increased level of digoxin and/or digitoxin when progesterone was added to the cultures. Stohs and Staba studies on the biotransformation of cardenolides by Digitalis cells and recognized that the glycosylation reaction occurred.

Among many studies, the biotransformation from digitoxin to digoxin using Digitalis lanata cells investigated by Reinhard and Alfermann is the most interesting approach in terms of commercial application since digoxin has a higher demand as a drug for heart diseases than digitoxin. It is advantageous that Digitalis leaves contain a larger amount of digitoxin which can be used as a substrate. It is a hydroxylation reaction at the 12 β-position of digitoxin, and Reinhard *et al.* found that β-methyldigitoxin was the most suitable substrate in this biotransformation as methyldigoxin is the major product.

VALEPOTRIATES

Plants in the Valerianaceae have been used as folk medicines. For example, Nardostachys jatamansi, Valeriana wallichii and V. officinalis L. var. angustifolia have been used in India, and N. chinensis has been employed in China for hundreds of years. Partrinia plants are also being used as sedative drugs in former Soviet Union. Although the active principles in these plants have not been identified, a group of compounds having biological activities such as sedative, tranquilization, cytotoxicity and antitumor activities were named "valepotriates". Thies synthesized a series of valepotriate derivatives and tested for biological activities. It has investigated plant tissue cultures for producing valepotriates because of their limited supply and uncertain availability. They induced callus tissues of nine different species of Valerianaceae on MS media and found that Fedia cornucopiae and V. locusta cells produced higher levels of the compounds than the intact plants. Isolation of cell lines resistant to trifluoroleucine and to nystatin, treatment with colchicine, cultivation of the cells in two-phase culture media and addition of several bioregulators were carried out intensively.

As valepotriates have monoterpene skeletons, L-leucine was considered as a precursor. In an isolated cell lines of V. wallichii resistant to trifluoroleucine, a leucine analog, but the yield of valepotriates in the cells was not increased although the intracellular level of leucine was increased. It was reported that fungi resistant to nystatin, a polyene antibiotic, produced a high level of steroids.

One of the strains of V. wallichii to resistant to nystatin increased the amount of valepotriates produced up to 3 times, which was 88 mg/g-d.w. The same group treated a suspension culture of V. wallichii with colchicine which was expected to induce polyploid cells. As a result, they found that the colchicine treated cells produced higher amount valepotriates than the respective untreated cultures. A two-phase culture provided by addition of RP-8 into the culture medium was effective in inducing secretion of the lipophilic compounds from the cells, and the total yield of valepotriates was substantially increased.

Since the valepotriate skeleton is of the iridoid nature, they added some plant bioregulators such as dimethyl-morpholinium-bromide, dimethyl-piperidinium-bromide, dimethyl-piperidinium-chloride as well as 2-(3,4-dichlorophenoxy)-triethylamine and 2-(3,5-diisopropylphenoxy)-triethylamine to cell suspension cultures of F. cornucopiae and V. wallichii. When they were employed in concentrations of 0.01 to 0.04 mmol during the early exponential growth stages of the cells, the levels of valepotriates were increased significantly.

Bibliography

A.K. Aditya and P. Haldar.: *Biodiversity and Environment*, Daya Publications, Delhi, 2000.

Arvind Kumar.: *Biodiversity and Conservation*, APH publications, Delhi, 2005.

Arvind Kumar.: *Biodiversity and Environment*, APH Publications, Delhi, 2004.

Ashish Dutta.: *Biodiversity and Ecosystem Conservation*, Kalpaz Publications, Delhi, 2001.

B D Joshi, C P M Tripathi and P C Joshi.: *Biodiversity and Environmental Management*, APH Publications, Delhi, 2009.

B N Pandey, Sadhana Desh Pande and B D Joshi.: *Biodiversity*, APH Publications, Delhi, 2007.

B.N. Pandey and M.K. Jyoti.: *Ecology and Environment*, APH Publications, Delhi, 2012.

B.N. Pandey, A.P. Sharma, P.N. Pandey, P.K. Katiha and K. Jaiswal.: *Biodiversity: Issues Threats and Conservation*, Narendra Publishing House, Delhi, 2012.

Bubul Das and Devashish Kar.: *Basic Limnology and Fish Biodiversity*, Manglam Publications, Delhi, 2011.

C.M. Chakarwarty.: *Ecology and Environmental Biology*, Swastik Publications, Delhi, 2007.

G. Madan Mohan.: *Ecology and Environment: Issues and Challenges*, Omega Publications, Delhi, 2008.

H K Pathak.: *Ecology and Environment: Current Perspectives*, Cyber Tech Publications, Delhi, 2009.

H.R. Singh and P. Nautiyal.: *Biodiversity and Ecology of Aquatic Environments*, Narendra Publications, Delhi, 2009.

John Spicer.: *Biodiversity*, Viva Books, Delhi, 2009.

K.S. Gulia.: *Ecology and Environment in the Himalayas*, Isha Books, Delhi, 2007.

Kumar Kartikeya.: *Biodiversity: Extinction and Conservation*, Aavishkar Publications, Delhi, 2005.

Lingaraj Patro.: *Biodiversity and Sustainable Livelihood*, Discovery Publications, Delhi, 2010.

M.M. Tiwari.: *Biodiversity and Environmental Conservation*, Monalisa Enterprises, 2011.

Mukesh Kumar, Rajan Kumar Gupta and G.S. Paliwal.: *Biodiversity: An Overview*, I.K. Books, Delhi, 2011.

N. Ramakrishnan.: *Biodiversity in Indian Scenarios*, Daya Publications, Delhi, 2006.

Nirmal Chandra Pradhan.: *Basics of Biodiversity*, Anmol Publications, Delhi, 2008.

P C Joshi and Namita Joshi.: *Biodiversity and Conservation*, APH Publications, Delhi, 2009.

P K Atre.: *Fish Genetics and Aquatic Environment*, Navyug Publications, Delhi, 2008.

P R Trivedi.: *Ecology and Environment*, APH Publications, Delhi, 2009.

P.C. Mishra, N. Behera, B.K. Senapati and B.C. Guru.: *Advances in Ecology and Environmental Sciences*, APH Publications, Delhi, 2012.

P.C. Trivedi.: *Biodiversity: Impact and Assessment*, Pointer Publications, Jaipur, 2009.

Pravin Chandra Trivedi.: *Biodiversity: Assessment and Conservation*, Agrobios Publications, Delhi, 2006.

R.N. Pati and Atul K. Jain.: *Biodiversity and Sustainable Development*, Sarup Publications, Delhi, 2010.

S.K. Singh.: *Dictionary of Ecology and Environment*, Commonwealth Publications, Delhi, 1997.

S.P. Pande and Leena S. Deshpande.: *Arsenic in Aquatic Environment*, Daya Publications, Delhi, 2010.

Shambhu Dhir.: *A Complete Book on Fish Genetics and Aquatic Environment*, Cyber Tech Publications, Delhi, 2010.

Supriyo Chakraborty.: *Biodiversity*, Pointer Publications, Delhi, 2004. T. Pullaiah and K. Jaganmohan Reddy.: *Biodiversity in India: Vol. 6*, Regency Publications, Delhi, 2013.

T. Pullaiah.: *Biodiversity in India: Vol. 1*, Regency Publications, Delhi, 2002.

T. Pullaiah.: *Biodiversity in India: Vol. III*, Regency Publications, Delhi, 2004.

T.R. Sahu and P.K. Sahu.: *Biodiversity and Sustainable Utilization of Biological Resources*, Scientific Publications, Chennai, 2004.

V.K. Prabhakar.: *Community Ecology and Environment*, Anmol Publications, Delhi, 2001.

Index